American Snakes

American Snakes
Sean P. Graham

FOREWORD BY RICK SHINE

JOHNS HOPKINS UNIVERSITY PRESS
BALTIMORE

© 2018 Johns Hopkins University Press
All rights reserved. Published 2018
Printed in China on acid-free paper

9 8 7 6 5 4 3 2 1

Johns Hopkins University Press
2715 North Charles Street
Baltimore, Maryland 21218-4363
www.press.jhu.edu

Library of Congress Cataloging-in-Publication Data
Names: Graham, Sean P., author.
Title: American snakes / Sean P. Graham.
Description: Baltimore : Johns Hopkins University Press, 2018. | Includes
    bibliographical references and index.
Identifiers: LCCN 2017004273| ISBN 9781421423593 (hardcover : alk. paper) |
    ISBN 9781421423609 (electronic) | ISBN 1421423596 (hardcover : alk. paper)
    | ISBN 142142360X (electronic)
Subjects: LCSH: Snakes—United States.
Classification: LCC QL666.06 G6815 2018 | DDC 597.96—dc23
    LC record available at https://lccn.loc.gov/2017004273

A catalog record for this book is available from the British Library.

Title page photograph by Bob Ferguson

*Special discounts are available for bulk purchases of this book. For more
information, please contact Special Sales at 410-516-6936 or special
sales@press.jhu.edu.*

Johns Hopkins University Press uses environmentally friendly book
materials, including recycled text paper that is composed of at least 30
percent post-consumer waste, whenever possible.

Contents

Foreword

Some kinds of animals are easy to observe, and it's not too difficult to work out what they are doing and why they are doing it. Almost anywhere in the world, children can get a good look at birds and bugs in their backyards—and perhaps even see them singing or feeding. But snakes are different. They don't make it easy for an observer. All you're likely to see is a thin body hurtling away from you into some kind of shelter—a hole in the ground, a bush, a tree—that you can't enter. And so, if you want to understand not only what kind of snake it was but also what it was doing when you interrupted it, you need help. You need to ask someone who has spent his life learning about the private lives of serpents.

There aren't many people in that category, but Sean Graham is one of them. Like most snake biologists, Sean's career represents a lifelong love affair with the animals that have inspired him ever since he was a small child. As a university professor, he studies the ecology and behavior of snakes, and gives lectures to students about those topics. And when he's not working, Sean travels to exotic locations to see even more snakes. As a result, he has accumulated a plentiful supply of firsthand experiences about snakes, and about their interactions with people. Those anecdotes enrich and enliven the authoritative information in this entertaining book.

American Snakes is about the day-to-day lives of these secretive animals. It might help you identify that snake you just saw, and it will definitely help you to understand the creature's biology. And you will marvel at the beauty of serpents through the spectacular photographs that adorn the pages. In a light-hearted and enthusiastic style, Sean explores the intimate world of snakes. He summarizes an enormous amount of scientific literature, but with an informal approach that makes it easy to follow. In the process, Sean has created a remarkably effective piece of public relations for a group of animals sorely in need of champions.

The unpopularity of snakes extends to many cultures—indeed, snakes are even less popular in Australia, my own home, than they are in North America. There's a simple reason for that difference; most of the snakes in Australia are venomous, whereas most American serpents are not. But even in Australia, snakes pose little real risk to people, and their danger is greatly exaggerated. I wrote a book in 1991 to provide a more scientifically accurate view of Australian snakes, and I'm delighted that, many years later, the same kind of book has now been written about the intensively studied snakes of North America.

American snakes have not been especially kind to me over the course of my career. I spent a few years in the United States soon after completing my doctorate, but my attempts to conduct field studies were a dismal failure. Attuned to Australian conditions, I simply didn't understand what American snakes were doing, where they were doing it, or how I could eavesdrop on their private lives. Defeated, I abandoned fieldwork and focused on laboratory studies and literature reviews until I returned to Australia and to the snakes that I understood better. If Sean's book had been available back then, I might have had more success.

Once, decades ago, I mistakenly seized a cottonmouth (Sean's favorite species), thinking it was a harmless watersnake. The incident occurred on the Texas-Oklahoma border, when I saw a snake swimming across a small stream and confidently took hold of it. That young cottonmouth wasn't to blame for the bite he gave me in return; he was only defending himself. As Sean's book reveals, snakes don't attack people. But this incident gave me a powerful and immediate lesson in my own fallibility in telling the difference between venomous and harmless snakes, and reminded me that snakes always have something new to teach us, even those of us considered to be experts. I spent a few days in the hospital dwelling on those lessons, and instead of feeling fearful or irate, I came out of the intensive care unit with a renewed appreciation of the diverse and fascinating nature of American snakes. The readers of this wonderful book can develop a similar appreciation without spending time in the hospital. I enthusiastically endorse that option, and heartily recommend this book.

RICK SHINE
Department of Biology
University of Sydney
Australia

Preface

And when those weapons are shown and extended for her defense, they appear weak and contemptible; but their wounds however small, are decisive and fatal. Conscious of this, she never wounds 'till she has generously given notice, even to her enemy, and cautioned him against the danger of treading on her.

Was I wrong, sir, in thinking this a strong picture of the temper and conduct of America?

An American Guesser (attributed to BENJAMIN FRANKLIN), 1775, on the rattlesnake as a symbol of America

The timber rattlesnake, one of our nation's first symbols. *Photograph by Noah Fields*

While conducting my graduate research on cottonmouths, I walked through the swamps three or four times a week. One swamp was behind a property owned by a man named Mike Dailey, the father of a childhood friend. Mike is a Florida native who grew up in the swamps and woods of the South, learned to hunt and fish at an early age, and is in all but one respect the archetypal Southern outdoorsman. He has trophy deer heads on the walls of his den, along with a stuffed largemouth bass and viciously posed taxidermy fox squirrels. There is a wooden plaque on his wall that commemorates when one of his hounds treed its first raccoon. Mike wears rubber Wellingtons and denim overalls while making the rounds around his property, and he has a nicely kept white beard. He would probably be flattered to be described as resembling Robert E. Lee, which he does. Once he stopped me on my way up the private drive to his house after I hadn't visited for many months, and he hailed me with his right hand up and the other resting on a .357 in its holster. He didn't remember my truck and wasn't taking any chances. Mike should despise snakes, but he doesn't.

Southerners are legendary snake haters. "The only good snake is a dead snake" is a common Southern phrase. Rattlesnake roundups—events for which thousands of rattlers are rounded up every year and butchered—are mostly Southern festivals. Attitudes toward snakes and other wildlife are terrible in the South. Why? Is it because there are more venomous species in the South? Perhaps, but Arizona has more venomous snake species than any other state (a bunch of rattlesnakes and one coralsnake), yet I've spoken with a firefighter in Tucson who underwent rattlesnake removal training. When people call the fire department about a rattler in their garage, the firefighters scoop them up in trashcans and release them *unharmed*. This would be unheard of in the South. Whatever the reason, Southerners usually hate snakes, and since Mike Dailey is as Southern as they come, he should hate snakes too. But Mike Dailey likes snakes. Hell, he *loves* them.

I don't know how exactly Mike ended up loving snakes, but he describes growing up catching them, much to the disapproval of his parents, who thought him strange for it. He used to swim with alligators when he was a child. He once had a pet eastern diamondback rattlesnake he would pick up barehanded. I would factor in an extra 30 minutes into my research trips when accessing my study site through Mike's property, because each time we'd shoot the breeze for a while about snakes. His house backed up to the felt-green floodplain of a Southern river that was full of snakes, the most common of which was my study animal: the venomous and dreaded cottonmouth. He was glad he had such a dense population of cottonmouths near his house. He said with pride that his swamp was where God put the first two. Can you believe this guy? He'd tell me about the snakes he'd seen lately, and maybe a funny story about some snake experience.

My favorite story he tells is about a snake he found while out hunting with a couple of buddies, neither of which knew much about snakes. He bet them a dollar that he could kill the snake with his urine. They took him up on the bet, and he proceeded to pee on the snake, which rolled over flat on its back, hung its mouth open and tongue out, and convincingly died. He collected the money from his friends, who stood slack-jawed in disbelief. The snake was an eastern hog-nosed snake—a species with a well-known death-feigning display that can be triggered by a predatory attack, a passing car, or in this case, an obscene act.

Mike wants to know everything about snakes. It is difficult to know whether he fell in love with snakes because he learned about them, or if he wanted to learn about them because he loved them. I would often tell him some things about my study, which involved the reproductive cycle of male cottonmouths. I was eager to observe courtship and mating in the population I studied, and once told him that snakes mate right after they shed their skin. His face showed surprise and asked me a question that stuck with me.

"Where do you find all this information about snakes? Is there a book I can read to learn all this stuff?"

I thought about it and reluctantly told Mike that I'd learned it all in scientific journals—literature that is typically only available at university libraries, and is dense, snooty stuff to read. It is off-limits to general readers, both because it is often unavailable at local libraries, and more importantly because it is written in the boring, scientific style of overeducated nerds. This literature is scattered across several decades, across dozens of journals, and across a variety of topics. I told him that the book he wants—a snake book that describes the everyday lives of American snakes—doesn't exist. Mike has lots of books about wildlife, but they are all the typical books available to those interested in snakes and other animals: field guides and reference books that are useful and informative, but not something you would read cover to cover. A readable summary of our snakes is needed for the American public, and the book in your hands is my attempt to deliver it.

No matter your background, or whether you find snakes beautiful or dreadful, snakes are inherently interesting. When I bump into locals at a fishing hole, one of the first topics to come up is the local snakes. When I tell people that I study snakes, they are quick to report a recent snake sighting or a colorful snake story. Television and Internet media have recently been fertile ground for unsubstantiated stories about giant, deadly, or disease-spreading snakes, and they run these stories because they sell. Outdoor magazines and *National Geographic* articles rarely fail to mention the teeming snakes present in some wild place they are covering. Snakes are among the most misunderstood and polarizing of the world's animals, yet the fascination they hold with the public cannot be denied, and that fascination has been growing steadily for decades.

This book is also about snake people. Sprinkled throughout the manuscript are short biographies about important snake biologists. All of them are interesting, dedicated, and smart people. After knowing many of them for years, and interviewing the rest for this book, I have learned what makes a snake biologist. Most grew up outside running around catching bugs, frogs, and other slimy things until they caught their first snake. For many, it was a garter-snake that started it all. For others, rarer and more local snakes were their first. For one, an afternoon hike with grandma revealed an eastern milksnake. For another, it was a prairie kingsnake in Peoria, Illinois. A California kingsnake inspired one English major to change paths and become a successful biologist.

The reason that I'm drawn to snakes is that they are quintessential under-dogs. Most of the snake biologists I spoke to said the same thing. When I was in elementary school, I wasn't a big kid, but when the class bully tried to pick on some hapless pencil neck, I'd usually try to get in the way. This never esca-lated into a fight, and it's a good thing, because I would've been pulverized. I've always looked out for underdogs. It's the way I've always been. Once when I was a kid, we spotted a watersnake up in a tree. The other kids readied to shoot it with a BB gun. Remembering the Greenpeace tactics I'd seen on TV, I stood in the line of fire.

Despite how interesting and ecologically important they are—and really, how downright gorgeous they are—snakes have been persecuted for crimes they are not responsible for. Few people like them in the same way they do the beloved little birds and cute furry little athletic mammals. Snakes are misun-derstood and mistreated, mostly owing to folklore and in large part to ancient religious stigma.

Snakes are important predatory animals and are found all over America—from the cottonmouth-infested swamps of the South to alpine meadows above 10,000 feet. They are found from the slick rock canyons of the Four Corners to the gleaming blue coast of California. Snakes are found underfoot in the rich humus of cove hardwood forests and overhead in the canopy of our oak hick-ory forests. They are found in our iconic deserts and rugged mountains as well as in our cities. You can see snakes in Hoboken, New Jersey, within sight of the Statue of Liberty. Snakes can be found under downed basketball backstops in Cincinnati, Ohio. And of course snakes are common in remote locations like the impenetrable no-man's-land of Okefenokee Swamp and the burning bottom of the Grand Canyon. Some snakes are widespread and found coast to coast, while others only just reach America, barely crossing our southern border with Mexico. Native snakes are not found in the forty-ninth and fifti-eth states, although one was recently found for the first time in Alaska, where it was probably introduced. Hawaii's waters are occasionally visited by the yellow-bellied seasnake, and it has been colonized by the globetrotting Brah-miny blindsnake. It may only be a matter of time before the invasive brown treesnake makes its way into a shipping crate to Hawaii, just as it colonized Guam, a US territory.

American snakes are as much a part of our rich and proud natural heritage as better-loved icons like redwoods, bald eagles, and grizzly bears. They have indeed been a part of our national symbolism, and before the bald eagle became our national symbol, a rattlesnake proudly graced an American flag, warning the British, "Don't Tread on Me." The flag has been recently readopted by the Tea Party movement. A snake even infiltrated the ranks of our Major League Baseball teams: along with the Baltimore Orioles and St. Louis Cardinals, there are now the Arizona Diamondbacks.

DONT TREAD ON ME

This flag was used to intimidate the English during the American Revolution.

While snakes are still reviled by many Americans, the time has finally come when most people are at least curious about them, and the number of people who are interested in them, respect them, and love them is growing at an astonishing pace. When I was growing up, I was the only kid in the class who wanted to hold the snakes when Okefenokee Joe—a local snake expert and enthusiast, and my first scientific hero—came to my school. The last time I did an outreach program with kids, I managed to get every single one to hold a snake.

This book offers a thorough explanation of the behavior, habits, and everyday lives of American snakes that is suitable for any reader. There are plenty of pretty pictures of the snakes doing their thing, and, when possible, I let the pictures speak for themselves, and try not to describe complicated behaviors with words. I've written it in the easiest and least technical style that I could. I have therefore purposefully avoided using scientific names, so I apologize to those connoisseurs who wanted them. I also followed the standard common name list published by the Society for the Study of Amphibians and Reptiles, which employs interesting word combinations like "cornsnake" and "gartersnake." My apologies to English teachers who might find these objectionable.

It's a proud celebration of a fascinating and diverse wildlife group, some of which are distinctly American. I hope this information will help quash some of our ignorance about snakes, leading to more snake lovers and fewer snake killers.

Mike, this book is for you.

Acknowledgments

First and foremost, I thank my wife, Crystal Kelehear, for proofreading this manuscript many times, for helpful suggestions, and for her excellent figures, which she stayed up until two o'clock in the morning the night before my deadline to finish. She has of course helped in additional innumerable ways and was patient with me while I worked on this. I love her and should give her at least six hugs a day. David Steen provided invaluable assistance in the form of his friendship, library login, and many helpful conversations and encouragements. I want to acknowledge Curtis Callaway, Noah Fields, and Tim Cota, photographers who not only provided outstanding photos but also went "on assignment" to get some of the shots I needed. The rest of the photographers also deserve special recognition for this beautifully illustrated book.

Thanks to all the snake biologists who agreed to interviews for this book. Special thanks to my mentors, who taught me everything I know about snakes that the snakes themselves did not teach me: Laine Giovanetto, Harry Greene, Craig Guyer, John Jensen, Gordon Schuett, Rick Shine, David Steen, and Dirk Stevenson. Thanks to J. D. Willson for providing many photos and helpful comments. Rick Shine deserves special mention for his encouragement and for his role in making this book a possibility; his book was the inspiration for mine. I also thank Vincent Burke and others at Johns Hopkins University Press for believing in me, for giving me a chance, and for their editorial guiding hands. My postdoctoral advisor, Tracy Langkilde, and department chair Chris Ritzi deserve mention for allowing me the time to develop and work on the book.

Finally, thanks to my parents, who indulged and encouraged my love of snakes from an early age.

American Snakes

1 · Introduction

I tell the Indians that I wish to spend some months in their country during the coming year and that I would like them to treat me as a friend . . . I tell them that all the great and good white men are anxious to know very many things, and that the greatest man is he who knows the most; that the white men want to know all about the mountains and the valleys, the rivers and the canyons, the beasts and birds and snakes.

JOHN WESLEY POWELL *The Canyons of the Colorado*

A Mamba in Utah

After working as a camp counselor one summer, I invited some of my coworkers to come with me on a float trip through the canyon country. They were a pair of South Africans who had no transportation but wanted to see the American West. So we piled into my parents' station wagon and headed for Utah, to canoe the Green River through Canyonlands National Park. The river there is

A striped whipsnake prowls the Great Basin Desert. *Photograph by Robert Hansen*

perfect for canoeing because it has none of the giant rapids like the Colorado River in the Grand Canyon. There are also no restrictions on access to the river, so there is no 10-year waiting list to run it. We paid a modest sum for shuttle service out of Moab, and they drove us and our canoes up to the put-in. The driver was a gruff, reticent, bearded old dude who, when pressed, told us he remembered seeing Edward Abbey strolling around Moab back in his park ranger days. He explained that Abbey was a gruff, reticent, bearded old dude. In five days, he'd send a pontoon to pick us up 114 miles downriver. We never saw anybody else.

Edward Abbey, the frank, sometimes profane, sometimes poignant, and always eloquent celebrator of the American West, took a trip down this same stretch of the Green River in 1980, on the eve of Ronald Reagan's election. Abbey had a copy of Thoreau with him, and he crafted a curious travelogue and critique of Thoreau's writing and life, concluding that we need more men like Thoreau and, of course, more wild rivers. Abbey was perhaps overly cautious about snakes but still sensed a kinship with them, concluding that "it is my duty as a park ranger to protect, preserve and defend all living things within the park boundaries, making no exceptions . . . I'm a humanist; I'd rather kill a *man* than a snake."

We floated lazily down the Green under a pounding August Utah sky, but when it got too hot we'd jump in the cool water. Occasionally we stopped and had mud-slinging fights along the tamarisk-lined banks, acting like a bunch of much younger boys. We climbed up red hoodoos greased with desert varnish, visited the rock signal towers of the ancestral Puebloans, and attempted to interpret their inscrutable petroglyphs. As we floated, the rocks slid past with the slow, determined pace of the river that was wearing them down—towers of sandstone perched atop hourglasses of pitted limestone—slapdash castles of rock in a maze of deeply red rock that was textured like rich chocolate.

On the first night, we camped in a copse of junipers across the river from a gleaming dune field. I examined a map and observed how remote our location was. There were 30 miles of desert between us and the nearest road. From that road it was probably an hour-and-a-half drive to Moab and the closest hospital. Any mistake would be paid for far out of proportion to the lightheartedness of our adventure. I had a strange moment of panic, something like a feeling opposite to claustrophobia, but with the same mounting anxiety. The other guys seemed fine, and by the next day on the river I was again comfortable and enjoying myself. We drifted among giant sandstone outcrops and around the long, aimless hairpin turns that are characteristic of Labyrinth Canyon. We floated dream-like past strange places with wonderfully poetic Old West names like Grand View Point, Horsethief Point, Horseshoe Canyon, Deadhorse Canyon, Hell Roaring Canyon, Bighorn Mesa, and Deadman Point. It was as if cowboy mapmakers had placed their own small talk into a Stetson, tossed in some landscape words for good measure, and then drew at random.

The river here carves some of the most breathtaking vistas of the West—when you stand at Grand View Point looking out from Canyonlands' Island in the Sky, the Green River is the sliver of brown water you see snaking its way down below.

At night, we'd brew up tea and tell all kinds of stories. Lacking a copy of Thoreau or Abbey, I instead read aloud from John Wesley Powell's account of the first float trip down the Green and Colorado Rivers, which took him down this stretch of the Green in 1869. Powell was a professor of little fame before his great expedition down the Colorado River through the Grand Canyon. He donated his arm for the Union cause at Shiloh, but he survived the Civil War with his lust for adventure intact. He turned to exploration, and after a few trips out west decided upon exploring the last part of America that remained unmapped. He organized an expedition with a handful of adventurers in small boats. This part of the voyage was easier than almost any other section; the Green and Colorado Rivers were mostly a harrowing series of rapids that had to be dangerously lined or strenuously portaged. In this canyon they had a respite from the crashing waves that they and their boats were wholly unprepared for, and they, like my friends and I, floated lazily along in safety. Of Labyrinth Canyon, Powell wrote, "The landscape everywhere, away from the river, is of rock—cliffs of rock, tables of rock, plateaus of rock, terraces of rock, crags of rock—ten thousand strangely-carved forms; rocks everywhere, and no vegetation, no soil, no sand. In long, gentle curves the river winds through these rocks."

After I read from Powell, Jerome obliged us with stories about his time in the South African military. Jerome looked a little bit like a young Mel Gibson: tan, bright eyed, with a great smile. He had been some serious Special Forces type and claimed that at one point during the bad times of apartheid he had Winnie Mandela in his sights during a protest but was never given the order to shoot. This is not to say that he shared the outlook of his superiors or his government at the time. He was progressive in his attitudes toward race relations in his country and supported the new Mandela government. He had impressed the other counselors at the ropes course with his head-first rappelling style, which he could use in combat and still have a hand available for shooting.

Wynand wore glasses and looked and talked like some sort of German villain from a World War II movie. He'd worked as a ranger on a private game reserve and told stories of rhinos pushing over Land Rovers. He explained that big game in Africa was big money, and that many farmers had converted their holdings to fenced-off reserves for tourists and, occasionally, some sustainable hunting. This sounded to me like a good idea that Americans could make use of for native wildlife. Both men had extensive experience in the African bush.

On the third day, we pulled off the river at lunchtime to explore a small side canyon. The river curled away at a 90-degree angle and brimmed off the side of a 1,000-foot wall of maroon sandstone before heading off to the south.

That wall of sandstone was cut cleanly in half—as if by a steak knife—by a small side creek that formed a box canyon. The creek mouth had its obligatory grove of cottonwoods and was merely a trickle. Before contributing directly to the Green River, it ended in a pile of mud cracks, under which scampered hundreds of recently metamorphosed red-spotted toads. I was photographing them when Wynand called me over. I walked up the white, book-layered silt banks of the creek and around a cottonwood, and found Wynand looking down at the creek bed. There, neat, clean tracks led up the box canyon.

"This is a big cat—leopard."

The individual tracks were about the size of softball, with a heart-shaped central palm adorned with four oval toes.

"Right," I said, "But it's a mountain lion. For sure."

We spent some time admiring the tracks. It was the first time I'd ever had such a close encounter with our ghost cat—the gorgeous, tawny killer of such diverse prey as white-tailed deer, sloths, desert bighorn, guanacos, mountain goats, agoutis, mule deer, and alligators. I felt the tight inside wall of a pugmark and felt each toe. The South Africans needed no help noticing there were no claw prints—the giveaway for the print of a cat rather than a wild dog—and the prints were the size of our hands, indicating this was no small cat. They were fresh, with smooth sides and no evidence of weathering. The mountain lion probably made them the previous night. Or that very morning. Or it was lying in the shade of the cottonwood, saw us approaching, and decided to flee up the box canyon five minutes ago. The cat could have been anywhere in that canyon, or up above on the walls, watching us. It was very exciting, mixed with just a pinch of disconcerting.

We fanned out along the canyon, hoping for a peek at the cat, or evidence of a kill, or even just to find a scat.

Then I saw the snake.

It seemed to appear out of nowhere, right out from under my feet, and darted toward the trunk of a cottonwood, which was grown over with shrubs. It was fast. I thought the guys would love to see this snake, so I dashed across the sand and dove for it, barely catching the tip of its tail as it disappeared into a shrub. My right hand was completely thrust into the shrub, and I was left flat on my belly, barely holding on. I was stretched so flat that I couldn't raise myself up or I'd lose my grip. I shouted for Jerome and Wynand. They appeared as the snake began inch-worming out of my grasp. Snakes are highly skilled at this escape maneuver, and unless you grasp them with a grip that can nearly break their vertebrae, they can wriggle their way out of your fingers like an accordion, a few scales at a time.

"Jerome, help me grab this snake!"

"What kind is it?

"I don't know."

"Then I'm not grabbing it."

"C'mon man, I know it's not venomous, look at it! It's long and skinny, I *promise* it's not venomous!"

"No."

I was losing my grip, and I was losing the snake, jerkily by the centimeter, as it found its way down a hole under the shrub.

"It's not venomous, just look at it!"

"I'm looking at it."

"It's not venomous—grab it!"

After pleading with him for what seemed like five minutes, with the tip of the snake's tail now pinned between the tips of two of my fingers, he reluctantly reached in and replaced my failing grip with his hand long enough for me to get up and reach in to grab it more firmly myself. After a few more seconds of finagling, I pulled out the snake. It looked like a gartersnake on steroids, about 3 feet long with cream stripes down its back, and big, bright nervous eyes. It had a pink belly. A striped whipsnake. It moved with rapid lunges in my hands, but in the sand it was a living ribbon. I was astonished how this veteran of the South African military could have shied at touching a skinny little American snake—completely harmless except to swift lizards, and entirely nonvenomous. I gave him a lot of grief for it.

Later that night, while pondering the situation further, I realized how entirely off base I was. In America, it's easy to identify venomous snakes with only a little practice. This is because nearly all of our venomous snakes belong to the same snake family; they're all pitvipers. All pitvipers have the same stout body plan—they're thick snakes with rough scales and distinct necks, giving them their legendary diamond-shaped heads. These features are mimicked with some degree of skill by some nonvenomous species, but if you mistake these for pitvipers, you've erred on the side of caution. The only skinny venomous snakes in the United States are the coralsnakes, which are brightly patterned with red-and-yellow bands that warn you not to harass them. Other nonvenomous snakes are extremely convincing mimics of coralsnakes, but here again, if you avoid these snakes assuming they are coralsnakes, you've erred on the side of caution. Even for the layperson, there needn't be an excuse for mistaking venomous snakes for nonvenomous ones. Of course, the opposite mistake is made all the time, and nonvenomous species are dispatched by the thousands each year because they are thought to be venomous species.

In many parts of the world, things are different. South Africa, where Jerome is from, has three or four families of venomous snakes, and they come in similar body plans and colors as the nonvenomous species. There are small, thin African burrowing snakes that can stab you from the side of their mouths with potent venom. African mambas and cobras are slim, bright-eyed, drably colored snakes not unlike American coachwhips and racers. In fact, they have a striped desert cobra that looks nearly identical to our desert whipsnake. But the African versions pack venom potent enough to send you to the hospital, or perhaps the morgue. Africa also has its share of camouflaged vipers that look like pitvipers in the United States. So, in Africa, you have to be a lot more careful when you go diving into a shrub to catch a snake. Sometimes you have to use a hand lens to observe minute scale differences in order to identify whether you have in hand a harmless snake, or one that can kill a man.

That evening, as the sun sunk low and orange over Canyonlands, we were camped on a long, parallel red rock shelf about 20 feet above the swirling brown water of the Green River. While enjoying our tea, I apologized to Jerome for questioning his courage, and thanked him heartily for assisting with the capture of such a fine desert snake. Somewhere out there in the dark, the mountain lion watched.

The Diversity of American Snakes

American snakes are placed into five major groups, most of which are harmless to man. Two families—the coralsnakes and seasnakes (Elapidae) and pitvipers (Viperidae)—are represented in America by a total of just a couple of dozen venomous species, all of which should be considered dangerous. The threadsnakes (Leptotyphlopidae) and boas (Boidae) are primitive and harmless snakes represented by only a few species in the western United States. The rest of America's snakes belong to the large and diverse snake family Colubridae. Because California and Arizona each have at least one representative from all five American snake families, they tie for having the most snake families of any American state.

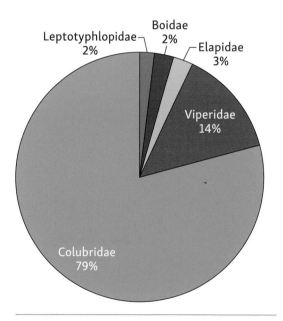

Percentage of US snake species in each of the five snake families.

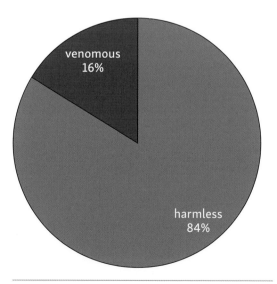

Percentage of dangerously venomous and harmless US snake species.

World distribution of US snake families. *A,* Leptotyphlopidae; *B,* Boidae; *C,* Elapidae; *D,* Viper-idae; *E,* Colubridae.

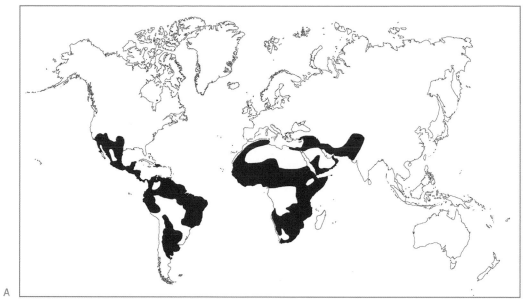

A

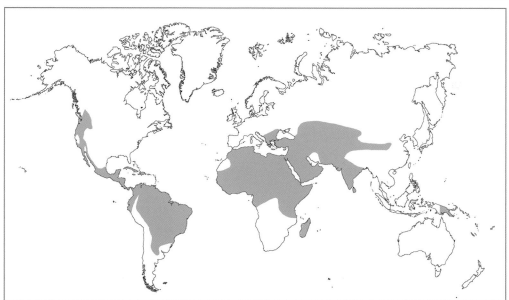

B

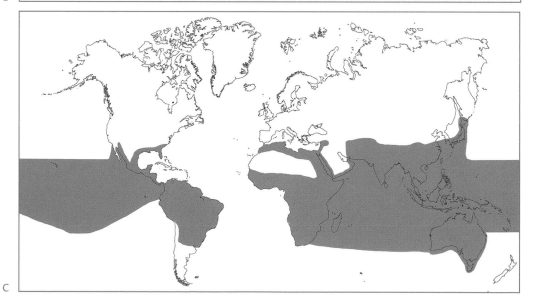

C

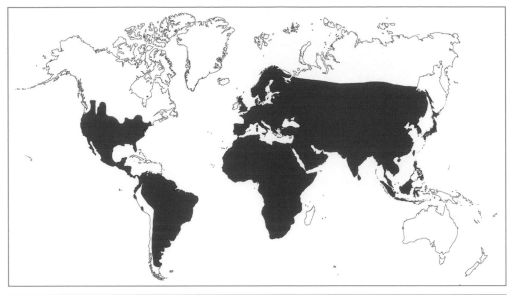

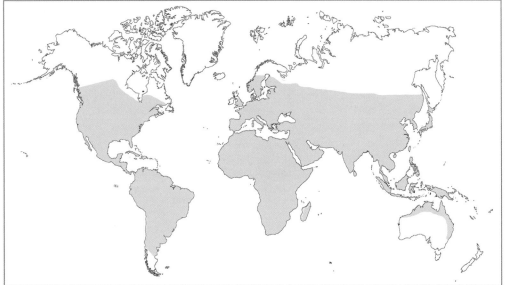

The rest of this chapter quickly outlines the origin and relationships of snakes in general, and explores the biogeography of American snakes.

Snakes Are Animals

Many people are surprised to learn that snakes are indeed animals and do not belong to some separate, yucky group of organisms. This may be because some people assume that "mammals" and "animals" are synonyms. When they go to the zoo to see animals, they expect to see giraffes, orangutans, zebras, lions, and perhaps an aardvark. Some might even think of birds as animals, too, and they would be right. Animals include a diverse array of organisms, and not all of them are warm and furry. Not all of them move. Fish are animals, but so are sponges. So are corals. So are sea squirts, which don't move and are more closely related to people than they are to insects. Insects are also animals. So are spiders, squid, tapeworms, centipedes, liver flukes, and leeches. For biologists, animals are simply multicellular organisms that ingest their food. Snakes are animals too.

Snakes Are Vertebrates

Many are also surprised to learn that snakes have backbones. Perhaps they think that snakes are some kind of worm. But snakes are vertebrates, and have closer kinship to birds, mammals, frogs, turtles, and salamanders than to invertebrates like leeches, worms, and centipedes. Snakes should be considered the ultimate vertebrates, because their skeleton is dominated by vertebrae. Most of their bones are backbones, and they have many more individual backbones (up to 350!) than any other vertebrate. Besides this distinction—which gives them their elongated appearance—snakes are not all that different from other vertebrates. In fact, the internal anatomy of a snake is hardly different from that of a man, something I'll discuss more in chapter 2. They have brains with the same structures as yours, only some are proportionately larger and some are proportionately smaller. They have a pair of kidneys, a pair of reproductive tracts, and a stomach and intestines with associated digestive glands like the liver, gall bladder, and pancreas, just like you and me.

When I worked at Zoo Atlanta, I would get calls from locals who wanted help identifying the various snakes that would turn up in their house. One caller described some kind of snake that had a diamond-shaped head and would shift in size from about a foot long to a few inches long, and left a slimy trail as it crawled. He was clearly concerned and wanted to know what kind of venomous snake it was. I told the caller the good news: not only was this venomous snake not venomous, it wasn't a snake. It turned out to be a kind of

Relationship of snakes to other terrestrial vertebrates. *Left to right*: amphibians (frogs, salamanders, and caecilians), mammals, turtles, squamates (lizards and snakes), crocodilians, and birds.

flatworm known as a land planarian. They were unintentionally introduced to North America in shipments of potted plants from Asia. These worms are about the same size and shape as a Thai noodle and shouldn't be mistaken for snakes.

Snakes Are Squamates

Snakes belong to a group of reptiles that are ectothermic (or "cold blooded"), have scales, and share certain features of skull anatomy. Snakes are essentially a successful group of limbless lizards that arose from within the same branch of the lizard family tree as Gila monsters and monitor lizards. They belong to a group of reptiles that is referred to as the squamates, which includes lizards and snakes but excludes turtles and crocodilians. Although turtles and crocodilians were once grouped along with lizards and snakes as "reptiles," this categorization has fallen out of favor as scientists have better understood the relationships between these animals. Squamates arose from ancestors that long ago would have looked a lot like a lizard, but because of their small size, their fossil record is incomplete. The history of snakes is full of extinct and living creatures with strange names that are unfamiliar to most people—lepidosaurs, scleroglossids, anguimorphs—and turtles and crocodilians are not as closely related to lizards and snakes as we once thought.

Grouping lizards and snakes to the exclusion of turtles and crocodilians highlights an important distinction: squamates are far more successful and numerous than the other "reptile" groups. With more than 6,000 species of lizards and 3,000 species of snakes, squamates are outnumbered only by birds in terms of terrestrial vertebrates living today. The diversity of squamate forms, colors, habits, and adaptations are comparable to the fabulous variety exhibited by birds and mammals, whereas turtles and crocodilians are far less diverse and are represented by far fewer species. And in the United States, snakes are by far the most numerous "reptiles."

The Origin of Snakes

The bizarre blindsnakes are the most primitive of all snakes, meaning they were the first to arise from lizard ancestors. They retain many characteristics that are shared with no other snakes. Although we know little about blindsnakes, what we do know of their form and habits contributes greatly to our understanding of snake origins. Because blindsnakes are completely ground dwelling and have reduced eyes, scientists believe that they arose from an ancient lizard—during a time in ancient geologic history when dinosaurs were around—that became elongate and limbless. The blindsnakes shed light on the origin of snakes, one of the most incredible evolutionary transitions known among vertebrates.

There is some controversy about the conditions under which snakes originally evolved from an ancestor with legs. But evolutionary biologists do at

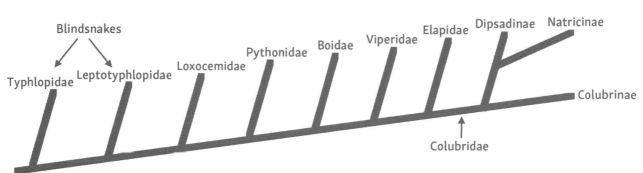

Relationship of American snakes to other groups of world snakes.

least agree on two things: one, the closest living lizard relatives to snakes are the fork-tongued "varanoid" lizards—which include the monitors and goannas of the Old World and the Gila monster of the American Southwest—and, two, losing legs is not at all special, evolutionarily speaking. Although there are no living varanoids that show limb reduction, there are fossil varanoids that do, and they demonstrate this was a likely first step.

It is estimated that leg reduction and total limblessness happened no fewer than 62 separate times in lizards. That number is worth stopping to think about. There are skinks, geckos, and dozens of other types of lizards that have elongated bodies with reduced or absent legs. Some groups of lizards include species that exemplify the complete range of limb reduction—from species with well-developed limbs to those with none at all—with all points in between represented by additional species.

Some of these lizards are found in America. There are four legless lizards in the Southeast (the glass lizards), and a few species of distantly related legless lizards in California. The unique Florida wormlizard, which lives in Florida and Georgia, is so heavily modified for subterranean life that its anatomical features are barely comparable to any other lizard group, making these strange lizards a challenge for taxonomists. The tiny Florida sand skink has puny limbs, a snake-like figure, and arose from within the common skink group in that region. So, if you include the snakes, in the United States alone there is evidence for at least five separate transitions to limb reduction or loss.

Snakes are essentially a heavily modified and species-rich group of legless lizards that diversified into the thousands of snake species present today. As such, snakes are by far the most successful group of legless lizards. Modifications of jaw structure further enhanced their success. Most other limbless lizards are insect eaters that have evolved limblessness solely for locomotion; they are able to efficiently slide through sand, crevices, or grass because they are elongate and move by curving side to side through lateral undulations. This is the most efficient way to move around in certain habitats. But the faces and jaws of these legless lizards are still like a lizard's, and they are best adapted for feeding on small insects. The eastern glass lizard is an excellent example. One can watch this snake-like lizard moving effortlessly through the grass, but it has eyelids, ear openings, and solid jaws just like any other lizard. Looking at its head is the easiest way to identify it as an imposter.

Instead, snakes have key adaptations associated with jaw flexibility that allow them to eat bigger prey, sometimes much bigger than their own head. This has allowed snakes to feed on a great diversity of animals. Unlike the numerous types of legless lizards, nearly all of which are insect eaters, snakes

A

B

Examples of animals with reduced or absent legs commonly mistaken for snakes. *A*, two-toed amphiuma, an eel-like amphibian with tiny legs (photograph by Curtis Callaway); *B*, Florida worm lizard, an amphisbaenian (photograph by Pierson Hill); *C*, sand skink, a type of lizard with tiny legs (photograph by Todd Pierson); *D*, a glass lizard, a type of legless lizard (photograph by the author). All of these are found within the United States, and none of them are snakes.

range from insect eaters to top predators in some ecosystems. Some snakes eat almost anything, while others will only eat one specific type of prey. There are snail eaters, egg eaters, fish eaters, crab eaters, crayfish eaters, bird eaters, bat eaters, rodent eaters, and even snake eaters. The ability to ambush or pursue diverse prey, combined with the ability to move smoothly through tight spots, is what makes snakes special.

There is a heated evolutionary debate about the conditions that produced snakes. Two opposing ideas of early snake evolution have been proposed: a marine-origins hypothesis and a terrestrial-origins hypothesis. There are fossils of ancient marine lizards (mosasaurs, one of the familiar sea monsters from dinosaur books) that are snake-like and have reduced limbs, and some early snake fossils come from marine deposits. There are also fossils of early snake-like lizards from terrestrial deposits, and considerable anatomical evidence that blindsnakes are most similar to the earliest snakes, which lends support to the terrestrial-origins hypothesis. Scientists who promote these opposing ideas trade shots every couple of years by publishing descriptions of new fossils or reinterpreting genetic information in fresh ways.

One of the more convincing lines of evidence for the terrestrial-origins hypothesis has to do with their eyes. Snake eyes are unique, having several novel features and functions not found in lizards (discussed in chap. 2). This suggests the rather remarkable explanation that snakes must have first went through a burrowing stage and lost their eyes, requiring them to evolve a brand-new pair essentially from scratch. A different analysis compared snake eyes to those of a variety of other vertebrate groups and found that snake eyes do have some things in common with aquatic vertebrates like fish, salamanders, and marine mammals. So they could have passed through a marine stage before evolving their unique eyes. Genetic studies on living squamates haven't been much help. The titles of two papers demonstrate the productive one-upmanship between the two sides: a paper called "Molecular Evidence for a Terrestrial Snake Origin" was published in 2004, and "Molecular Evidence and Marine Snake Origins" was published the following year.

Most early snake fossils have come from terrestrial deposits, and the marine fossils that so compellingly linked snakes to mosasaurs are now interpreted as advanced snakes similar to boas and pythons. So, for now, the evidence is leaning toward a burrowing snake ancestor. In fact, the snake currently considered to have emerged earliest was discovered right here in the United States. The snake is so old it was discovered accidentally by paleontologists looking for dinosaur bones. This small burrowing snake was first described in 1892 by Othniel C. Marsh, a paleontologist who took scientific controversies much further than the factions currently arguing over the origin of snakes. He was at war with rival paleontologist Edward D. Cope. The two sent each other nasty letters, scathing reviews, and gave their field crews strict orders to spy on, steal from, and interfere with their rivals at digs. The field crews carried weapons and sometimes literally traded shots because, after all, they were digging up dinosaurs (and primitive snakes) in the Old West. *Coniophis* was described by Marsh from a single vertebra, but from that fossil it was hard to tell exactly what kind of snake it was. Much more material came back from the digs and sat unexamined in museums for decades. In 2012, scientists found fragments of multiple bones that belonged to the snake filed away among dusty museum shelves. When they examined them, they determined that the skull features placed *Coniophis* directly between lizards and snakes. The facts that *Coniophis* was found in terrestrial deposits, looked a lot like a modern sand boa, and had other skeletal features associated with burrowing lend support to the terrestrial-origin hypothesis.

The current consensus is that snakes started as a varanoid experiment with an underground existence, followed by adaptations for termite feeding, which produced blindsnakes. More familiar snakes are the result of a move back above the ground, which required a new set of eyes and an unparalleled set of jaws, which led to the eventual evolution of rattlesnakes, mambas, anacondas, taipans, and seasnakes—all the diverse and more familiar serpents that have spread onto most continents and into the world's oceans. If snakes had never ventured back above the ground, we'd probably simply refer to blindsnakes as just another small group of interesting termite-eating, legless lizards.

Blindsnakes, Leptotyphlopidae

There are about 300 blindsnake species worldwide, and most are found in tropical and desert regions. Only a few species occur America naturally (the threadsnakes), and they are all found in the hot deserts of the Southwest. One tiny species, the Brahminy blindsnake, has been introduced by humans throughout the world in the soil of potted plants, and it has made its way into Florida, Georgia, and Hawaii. As a clonal species, this tiny snake is an exceptional colonist. A new population begins with only one individual laying eggs, because each new hatchling is a daughter identical to her mother.

A threadsnake, one of three kinds of blindsnakes found in the American Southwest. *Photograph by Noah Fields*

As the name implies, blindsnakes have eyes reduced to small spots of pigment partially concealed by scales. Their eyes are probably only useful for determining whether it is night or day and perhaps for maintaining the snakes' internal seasonal clock. Blindsnakes have some primitive features not found in other snakes. They do not have the enlarged belly scales typical of other snakes, and instead all of their scales are the same size. Internally they have a residual pelvis. They still have a pair of lungs, whereas more modern snakes have lost one. Most blindsnakes eat ants and termites, and live among their prey underground or in termite mounds.

The Boas, Boidae

Boas are another ancestral group of snakes, and they demonstrate early evolutionary experiments in gigantism and jaw flexibility. Boas have some features that are intermediate between the primitive blindsnakes and more advanced snakes. Boas have enlarged belly scales compared to blindsnakes, but these scales are still small compared to most other snakes. Males of some species retain pelvic bones useful for clasping females during courtship, their so-called

A rubber boa, one of the boas native to the United States. *Photograph by Marisa Ishimatsu*

spurs. The relationship between boas and other snakes has been controversial. Some scientists have shown that boas and pythons are each other's closest relatives and should be considered a single family, while others have argued that boas are more closely related to other, more obscure snakes than to pythons. Regardless, the boas and pythons are only represented by a few dozen species worldwide, yet they have managed to become the planet's largest snakes and top predators in some tropical regions. They are also among the most familiar and feared of all snakes, and they are the only snakes known to have consumed humans. Despite their fictional roles in movies as man eaters, boas and pythons are harmless to humans in terms of venom, and most species are quite a lot smaller than the anacondas of South America and reticulated pythons of Asia. Your chances of being consumed by a large boid are far less than your chances of being murdered and consumed by another human.

America is home to only four boas, all of which belong to a small subgroup (the Erycinae) found in semiarid areas in North America, Europe, Asia, and North Africa. The rubber boas are found in varied and surprisingly cold habitats in the West and range north into Canada. The rosy boas are desert and chaparral dwellers found in California and Arizona.

The United States is also becoming home to exotic boas and pythons because of intentional and unintentional pet releases. Most Americans are probably now aware of the unfortunate introduction of Burmese pythons to south Florida. Boa constrictors and African rock pythons are now being found regularly there as well. Green anacondas too occasionally turn up. These invasive snakes are covered in more detail in chapter 10.

Colubridae

Most harmless American snakes belong to the large, worldwide family Colubridae, which includes thousands of species, some of which are not closely related to each other. For this reason, prudent calls have been made to split this unwieldy group into several smaller families. For convenience, I choose to treat them as one group following the research of Alex Pyron and his collaborators. American colubrids can be sorted into three smaller subgroups, each with their own independent history. The natricinae are a successful subfamily of small- to medium-sized snakes that often have keeled scales, giving them a rough appearance. All American species also share the reproductive adaptation of viviparity, meaning they give birth to live young. The family contains some of the most familiar harmless snakes, such as gartersnakes and ribbonsnakes, which are also some of our most well-studied species. Natricines arose in Asia, spreading from there to Europe, Africa, North America, and northern Australia.

A California red-sided gartersnake, a brightly colored natricine and member of the colubrids. *Photograph by Marisa Ishimatsu*

American natricines occupy semiaquatic and fully aquatic niches, with some species practicing an adaptable lifestyle and others having narrow feeding habits and habitats. Examples of the first strategy are the common gartersnakes and common watersnakes, which live in diverse habitats and feed on a wide variety of prey, depending on what the other snakes in the neighborhood eat. Examples of the second strategy include queensnakes, which feed only on recently molted crayfish. Natricines often make up a considerable component of vertebrate biomass in wetlands, with thousands of watersnakes occupying the swamps and rivers of the South, and hundreds of secretive earthsnakes, red-bellied snakes, and brownsnakes living in the leaf litter of eastern forests. Natricines are often the only snakes present in some American landscapes, and can be found along narrow watercourses in the desert Southwest and alpine meadows in the Rockies where no other snakes are present. Some species also thrive in cities, requiring only vacant lots to live, feed, and breed. Gartersnakes, ribbonsnakes, lined snakes, Kirtland's snakes, and Dekay's brownsnakes are some of the species encountered in our urban environments, and for this reason they are sometimes referred to as "city snakes." The common gartersnake is also our northernmost species, occurring well north into Canada. In such high latitudes, these snakes converge on the same rocky dens for hibernation and occur locally in outrageous numbers.

Dipsadines include some of our most common species and also some of the rarest and most highly prized species—many are shiny snakes so gaudily colored they could have spawned from the imagination of a child. A few have vertically elliptical pupils—like a cat's eye—a rare feature for harmless snakes. Many also have sharp tips to their tail, the function of which are unknown. It

A rainbow snake, perhaps our most coveted and gaudily patterned snake. *Photograph by J. D. Willson*

is a largely tropical family of snakes, and a minor but interesting addition to our American snake fauna. The best way to see an assortment of dipsadines is to travel south into Central America, perhaps to Costa Rica, and walk on trails or along roads at night. There they can be found in great variety and abundance, eating such exotic prey as snails, frog eggs, and highly toxic dart frogs. Some of these dipsadines have their northernmost populations in the United States and are otherwise much more common south of the border. The most ancient dipsadines are strange Asian species found near hot springs, so these snakes appear to have independently colonized North America from the natricines and subsequently diversified before moving south into the tropics. Many are among the more common and familiar eastern snakes, and it is not well known nor advertised to the lay public that they are technically venomous, albeit not to humans. In most species, the venom apparatus is so weak that there is still debate over whether they are truly venomous. Even the familiar ringneck snake, one of the most widespread snakes in the country, has tiny rear fangs that are harmless to humans but lethal to their small vertebrate prey. Most dipsadines use the venom to kill select prey items. Mudsnakes only eat large eel-like salamanders, for example, and rainbow snakes eat only eels. Hog-nosed snakes are toad specialists, and in captivity they will usually only accept alternative food if it has been laced with toad scent.

The Colubrinae is a worldwide subfamily of similar snakes with fairly slim body proportions. They are also diverse in Asia and are among the more common snakes in Europe, Africa, northern Australia, and the Americas. Many Eurasian species have similar habits and look like American snakes, so they appear to be closely related to them. Instead, most American colubrines are more closely related to their dissimilar American cousins than they are to the Eurasian species that they most closely resemble. For example, pinesnakes, gophersnakes, and ratsnakes are all fairly closely related, but American ratsnakes are not closely related to the "ratsnakes" of Eurasia.

Many colubrines prey on small vertebrates, and other colubrines have fairly specialized diets, with some species eating only centipedes (the crowned snakes) and others feeding exclusively on other snakes; for example, the short-tailed kingsnake of Florida feeds almost exclusively on a single snake species. The colubrines include many small species, but also the indigo snake, which at ~3 m long is our largest snake. The long, skinny shape of colubrines makes some species, such as racers and coachwhips, capable of crawling quickly, and likewise our arboreal (tree-dwelling) snakes are with few exceptions colubrines. A large diversity of arid-land colubrines probably originated in the deserts and associated highlands of Mexico, and therefore the Southwest contains a diverse group of desert colubrines. Like dipsadines and some Asian natricines, some colubrine species have small rear fangs with associated venom glands, but these species are harmless to humans.

The Coralsnakes, Elapidae

Elapids are a diverse and successful group of front-fanged, highly venomous species. All share a venom delivery system that includes small, fixed fangs on the front of their upper jaw, and potent venom that is usually neurotoxic. Africa, Asia, and Australia all have large numbers of dangerous elapids familiar to anyone who watches nature shows on TV—they include cobras, kraits, mambas, taipans, and death adders. Far less familiar are the small, colorful, and

Frank Burbrink

Punk Taxonomist

Arianna Kuhn

"I know, I know," Frank explained into the microphone, the groans of old-school snake biologists nearly drowning him out. This is how I remember him at a scientific conference years ago—standing at the podium, sweating, explaining his branching diagrams. His genetic research showed ratsnakes in North America are more closely related to other North American snakes than they are to ratsnakes in Eurasia. They do not form a natural group, so the entrenched name *Elaphe* would not work for North American ratsnakes. He suggested instead the perfectly valid and quite strapping name *Pantherophis*, which literally means "panther-snake." Still: groans. Frank continued, occasionally reassuring the audience with another "I know, I know," as he dismantled decades of taxonomic stability.

Frank's appearance and attitude probably encouraged the ire of the old guard: he wears Buddy Holly glasses and is covered in tattoos. At conferences he'd be wearing combat boots, field pants, and a large, smelly snake wrapped around his arm. He's loud and animated and isn't afraid to tell it like it is. He's also funny, intelligent, and warm. I liked Frank right away.

His research group created headaches for snake taxonomy, which has been in a state of flux for over two decades. Names we all grew up with have changed at a moment's notice. Subspecies have been elevated to full species while others have been discarded as irrelevant. Species have been carved from what at first glance seemed to be the same thing. Frank has been in the thick of all these new changes, so much so that you'll hear snake biologists refer to some snakes as "Burbrink species." But it's not as if Frank set out to cause trouble. He's only interested in testing ideas and increasing our understanding of the evolution and relationships of snakes, and he "has no problem kicking the legs out from under my own stuff." Frank determines evolutionary relationships based on genetic data; if new data become available, the story can change. This is how science works, and perhaps the only thing unusual about the recent changes is how rapidly they occur. Today, within days of a new paper coming out, the scientific name of a species can change.

Frank doesn't understand how he came to be a pariah. He never understood why people got so upset, referring to this as "the burden of heritage." He explained, "in ecology it's different"; if someone tests and falsifies an old idea, "nobody goes bananas."

I've seen the tide turn in favor of Frank's work. The taxonomic changes he supports are now quickly adopted. His students, including Alex Pyron and Sara Ruane, have become some of the most productive and influential snake systematists out there. Now Frank is curator of herpetology at the American Museum of Natural History in New York, which is among the most prestigious and influential positions in herpetology. The grumbling has stopped. Rock 'n' roll.

A harlequin coralsnake, which advertises its deadliness with bright colors. *Photograph by J. D. Willson*

secretive Asian coralsnakes, which are the ancestors of the American coral-snakes. Most Asian coralsnakes have fewer bands than American coralsnakes, and some have only a neck ring or colored bands on the belly. Some even have stripes. A few bizarre species have enlarged venom glands that stretch past their heads and well into their necks. The function of these enormous venom glands, and most of the biology of Asian coralsnakes, is unknown. But we do know that this small tropical Asian group sent a representative to the Western Hemisphere long ago, and this led to a successful adaptive radiation of doz-ens of coralsnakes in Central and South America. The United States has three species, one of which (the Sonoran coralsnake) is the most ancestral of all the Western Hemisphere coralsnakes. The harlequin and Texas coralsnakes are fairly closely related and were once considered subspecies of the same species.

Coralsnakes eat mostly small snakes and lizards, and seem to occur primar-ily where high densities of their favorite prey (small snakes) exist. Although they are not well represented by large numbers of species in the United States, and they are fairly uncommon and secretive, coralsnakes have certainly left their mark on the psyche of predators throughout the regions where they occur, including humans. Coralsnakes have alternating red, yellow, and black bands, and their presence and toxicity have led to the development of a num-ber of similarly patterned mimics—harmless snakes that are not venomous but benefit from being mistaken as coralsnakes by predators. The trick is so convincing that children in the South are taught a fairly elaborate rhyme early in life: "Red on yellow, kill a fellow, red on black, venom lack." Although I was taught this rhyme as a child, I've rarely had to use it; I've encountered only a handful of the mimics in my explorations of the South, and I have never seen an American coralsnake alive.

The Seasnake

A single representative of this successful group of marine snakes occasionally reaches the southwestern shores of California and the Hawaiian Islands. The yellow-bellied seasnake is the only open-ocean snake; most seasnakes are found along the nearshore coastal shelves and estuaries of Asia and Australia. Seasnakes are essentially marine elapids, specifically Australian elapids. There are so many elapids in Australia that they even diversified into the warm, shallow seas of Australasia. The pelagic seasnake became completely cut off from the land and its influence, and it cruises along ocean currents throughout the limitless blue Pacific, seeking the churning white foam of productive current

A yellow-bellied seasnake, an open-ocean dweller that occasionally visits the shores of Hawaii and southernmost California.
Photograph by Will Flaxington

confluences. As marine elapids, seasnakes have some of the most potent venoms of any snakes in the world. Fortunately, they are rather docile and inoffensive, and rarely bite people.

The Pitvipers, Viperidae

Vipers are found worldwide from Sweden to southern Africa through Asia, but skip Australia, which is so full of elapids that, in the absence of vipers, an elapid—the death adder—independently evolved a viper-like body and behavioral repertoire. Vipers invaded North and South America, like many snake groups, from Asia. The Old World vipers include several lineages of "true vipers," or those without pits. Large areas of Asia also contain pitvipers, which are also found throughout the Western Hemisphere. All vipers share a sophisticated venom delivery system that includes large, hollow fangs that hang from the bottom of a rotating upper jaw, such that during a strike the fang hinges out like a switchblade. Pitvipers have an additional modification on their face: a pair of heat-sensitive organs between the eye and nostril, called the loreal pit. This amazing adaptation is the most advanced heat sensory system known in vertebrates, and its function will be discussed in some detail in chapter 2.

America boasts a modest diversity of pitvipers, and its position as a waypoint for pitvipers between Asia and the New World tropics offers representative diversity. The American "moccasins"—copperheads and cottonmouths in the genus *Agkistrodon*—are the most ancestral American pitvipers and are probably the most similar to the first pitvipers that came from Asia. Three species of pygmy rattlesnakes are found in the United States. These are the most ancient of the rattlesnakes, and they probably resemble what the early rattle-

A Grand Canyon rattlesnake, a variety of the more widespread western rattlesnake found only within our iconic Grand Canyon. *Photograph by Timothy A. Cota*

snakes looked and acted like. They have much to tell us about the evolutionary origin of the rattle. This question is a favorite of pitviper biologists, a proud and masculine bunch who aren't afraid to get in heated verbal arguments during question-and-answer sessions at scientific conferences—especially when the topic is the origin of the rattle.

Three camps have emerged, each supporting its own idea of the rattle's origin, and the debate is exacerbated by the fact that none of them have had much success testing any hypothesis or designing any experiment to support their view. The arguments are mostly rhetorical but nonetheless interesting. The first idea is that rattles first developed as an early warning deterrent for hoofed animals so the snakes wouldn't be trampled, and later became the specialized antipredatory device they are now. This was later modified into the hypothesis that the early rattle evolved to warn small predatory mammals. The second idea is that the rattle first evolved to distract attention away from the snake's head during a predatory encounter. This is probably why rattling in non-rattlesnakes—which is quite common—is adaptive. The first rattles were perhaps even better at distracting predators and later became an effective warning device. The third idea is that rattles first developed as an effective lure for prey, later becoming the antipredatory device they are now. Support for this last idea comes from observations of the most primitive rattlesnakes—the massasauga and pygmy rattlesnake—which have tiny rattles that can barely be heard. This suggests that the early rattle couldn't be effective for warning anything. On the other hand, these small rattlesnakes have young with brightly colored tails that are used to lure prey. Most pitvipers have this ability when they are babies, and primitive rattlesnakes may have had a feeding advantage by buzzing a little insect-like rattle that over time became bigger. Later the rattle became an effective warning system.

All we know for certain is that the rattle did evolve, and rattlesnakes, possibly in large part because of this unique warning system, have diversified greatly in Mexico, and a few species even range to South America. Most rattlesnake diversity is a result of the presence of what are in essence Mexican species reaching their northernmost extent in our desert Southwest. The United States is also home to the world's largest rattlesnake species: the eastern diamondback of the South.

Pitvipers vary in their feeding from extreme dietary generalization to fairly specialized diets of mostly warm-blooded prey. For example, the cottonmouth may have the least choosy dietary preferences of any snake in the world; nearly every possible living thing, and many a dead thing, has been found in the belly of the cottonmouth. On the other end of the extreme is the eastern diamondback, which when fully grown eats mostly rabbits.

The venom of pitvipers is intended for killing their prey, which can be large and dangerous. Imagine attempting to eat a live rabbit without using your hands. Pitviper bites induce painful swelling at the very least. Bad bites frequently cause gruesome damage, and in certain cases a most unwanted outcome: death. Yet all American pitvipers are also extremely mild mannered and nonconfrontational until approached or handled. Legitimate snakebites (e.g., those in which the victim was unaware of the presence of the snake until bitten) in our country are becoming extremely rare, and in most cases, victims of snakebites were asking for them. We will examine the risk of snakebite more in chapter 9.

All American pitvipers are viviparous, giving birth to small numbers of large young. They also have an additional, surprising behavior: pitvipers are ever so briefly good mothers to their newborn babies. This astonishing fact will be discussed in more detail in chapter 5.

Truly American Snakes

Our five snake families bestow America with a magnificent snake fauna, with representation of some of the most primitive and advanced snake groups. Species diversity centers on the southeastern United States, where large numbers of colubrines and natricines occupy the diverse habitats of the warm and humid South. Evolutionary diversity is centered more in the Southwest, where primitive groups like boids and blindsnakes have representatives, and tropical dipsadines and colubrines are better represented. Biogeographically, the American snake fauna has its closest affinity with that of Asia, which was the ancient source for our elapids, dipsadines, colubrines, natricines, and viperids. Land bridges between North America and Asia allowed frequent exchange between these continents in ancient times, the most recent of which allowed migration of people to the Americas for the first time. The migrations of Asian snakes occurred much earlier, during Eocene (50 million years ago) and Miocene (15 million years ago) times. America has been the evolutionary epicenter for modest radiations of watersnakes, gartersnakes, and rattlesnakes, as well as what I'll call colubrid constrictors.

Each of these species groups should be considered truly American because they have a large number of species found across the United States and their diversity is centered here. They also contain species that range across the country and have diversified into local varieties. Watersnakes of some sort or another range over much of the United States and become diverse and numerous in the South. Within this radiation are the crayfish snakes, which are small olive watersnakes that have developed a select taste for crawdads and eat practically nothing else. Both the southern and common watersnakes have wide ranges across the eastern United States, and they have localized habits and color patterns that currently distinguish subspecies. The common watersnake comes in interesting island forms, such as the federally protected Lake Erie watersnake, a dark, plain form found around an archipelago in Lake Erie near Toledo. There is even an Outer Banks watersnake, a local variant that vacations off the North Carolina coast.

Alex Pyron
Recovering the Snake Tree of Life

When I was in college and still unsure of what to do with my life, I went on a field trip to the mountains of north Georgia to catch hellbenders—giant salamanders found in trout streams throughout the Appalachians. I was there to help the state herpetologist, who had invited a few others to lend a hand as well. We waded into the clear mountain stream and caught a handful of the slimy salamanders, which are so ugly they have a roguish charm. We even discovered a nest, which in this species is attended by the adult male. The gelatinous eggs were the size of eyeballs, the mass of them bigger than a dinner plate and unfurling downstream from under a rock.

One of the professors brought along what I assumed was a local middle school kid, surely one who'd developed an incurable passion for amphibians and reptiles, whose parents begged the professor to let him come along. He busily shuffled around the creek edge catching salamanders. But this kid was different. When I got a little closer and overheard the dialog between him and the professor, I realized he was spouting off technical terms and scientific names like it was old hat. After joining him to catch salamanders for a minute or two, I slipped into my encouraging camp counselor persona.

"Hey man, you're pretty handy with those scientific names for a kid your age," I said somewhat condescendingly, but I meant it as a compliment. I could have saved it.

Less than 10 years later, Alex Pyron would graduate with a PhD at 21 years old and is now well on his way to mapping out the family tree of snakes. He completed his doctorate and had a faculty job before I did, despite my considerable head start. His dissertation research determined the evolutionary relationships of kingsnakes and resolved their rather difficult taxonomy. He has accumulated samples of a few thousand snake species, and his genetic analyses are the most comprehensive ever attempted. I relied heavily on them for my interpretation of the evolutionary history of American snakes. To obtain genetic samples, Alex collaborates with snake biologists from all over the world and travels to far-flung localities like Sri Lanka and Madagascar in his quest to reveal the snake tree of life. To make the situation even more unbearable, Alex is not some kind of jittery nerd. In addition to being intelligent, he's an outgoing guy you can talk to like anybody else. To top it all off, he looks like Leonardo DiCaprio—although one supposes his dating opportunities are limited, given that Professor Pyron's available playing field is populated by students his own age.

Alex is one of those rare geniuses you hear about—a real child genius—who eventually goes on to win the Nobel Prize in astrophysics. We're lucky that he never shook his childhood interest in snakes.

Examples of snakes belonging to distinctly American groups. *A*, mangrove saltmarsh snake, a member of the natricine watersnakes (photograph by Pierson Hill); *B*, blue-striped ribbonsnake, a member of the diverse gartersnakes (photograph by Pierson Hill); *C*, twin-spotted rattlesnake, a southwestern mountain rattlesnake (photograph by Timothy A. Cota); *D*, green ratsnake, a colubrid constrictor also found in the desert Southwest (photograph by Frankie Casalenuevo).

Perhaps more than any other group, the gartersnakes have developed into uniquely American species. A large number of gartersnakes occur in the Midwest, and up to four species can be found in some areas, like northwestern Ohio. Some of these are worm-eating gartersnakes that have the smallest distributions of any of our snakes. With eight species, including some found nowhere else, California is another gartersnake hotbed. The common gartersnake is the widespread, locally variable representative of this group. It is found in endless varieties of color and geography: from the yellow-striped eastern gartersnake of the eastern meadows to the red-sided and red-spotted forms of the Great Plains and Pacific Northwest. Some of these have such small ranges they have taken the names of great American cities. There is the stunningly attractive San Francisco gartersnake and even a checkered Chicago gartersnake. What could be more American than a Chicago gartersnake?

Rattlesnakes too should be considered distinctly American, although Mexico is the true center of rattlesnake diversity. Many of our rattlesnakes have ranges that continue far south of the border. But we do have one wide-ranging rattler that has diverged into numerous local geographical varieties, and these are distinctly American. The western rattlesnake (*Crotalus viridis*) was once considered a single species with over a dozen localized forms, which compose a quilt-like distribution centered on the Colorado Plateau. There has been a recent push to recognize all of these as separate species, which would have the net effect of doubling the number of venomous snake species known in the United States. If this prospect alarms you, I can put you in touch with those responsible.

Whether we consider them local geographical variants or separate species makes no difference—these snakes are clearly evolving into distinctive-looking rattlers with their own personalities. These include the Arizona black rattlesnake of the Mogollon Rim; the brownish-yellow Pacific rattlesnakes of California's scrubby beaches and Coast Ranges; the common, greenish prairie rattlesnake of the Shortgrass Prairies; and the tiny, washed-out midget faded rattlesnake of the canyon country. There is even a tan-pink rattlesnake that is found only in the Grand Canyon.

The colubrid constrictors include a diverse array of medium to large, closely related colubrids that constrict prey. This group includes the king-, rat-, corn-, milk-, fox-, bull-, pine-, and gophersnakes. There is a Louisiana pinesnake, an endangered species found only in the fragmented pinelands of northwestern Louisiana and the Texas Cajun country. There is also a black pinesnake (also endangered) of the southern pine region, the bullsnake of the Great Plains, and gophersnake of California and the Great Basin. They are all large constrictors with a close kinship. Kingsnakes range across the country in confusing variations of a black-and-white banded theme. There is a wide variety of milksnakes, from the gorgeous California mountain kingsnake to the rather plain eastern milksnake. Like other truly American snakes, this radiation includes a profusion of locally adapted species with distributions centered in the United States, and some wide-ranging species that exhibit geographical variation within our country.

These local populations represent snakes that are actively evolving into new species. They belong to groups with a number of closely related species, and each of these includes a species that is widespread in America and has begun developing local adaptations. These local flavors are in the process of speciation, but they have not diverged quite as much as the other species in the radiation. For some biologists, it makes sense to call these local variants subspecies. For others, it is better to go ahead and call them species. The fact that authorities argue about what a species is and isn't—how far diverged they have to be, or how reproductively isolated they must become—is to me one of the finest proofs of evolution. If evolution did not occur—if all species were simply dropped from the sky to Earth—then it would be far easier to decide what's a species and what's not. Instead, it is not this obvious, and that is exactly what you'd predict if species diverge gradually from their ancestors. Some lineages of snakes are in the process of accumulating new adaptations that will make them more different from their relatives, and others have already diverged considerably. Some have close relatives they look similar to, and others have no close relatives because they've all gone extinct.

American snakes are fine examples of this process of speciation in action. They have developed into unique curiosities that are as American as our local foods and cultures, our beloved landscapes, and our distinctive fauna. Some have spectacularly American names and are tied to places so American you should find them depicted on a postage stamp.

So stop in at a roadside taco stand in Los Angeles, and within a few hours you could be walking among shining oak savannahs and spot a yellow-brown-blotched southern Pacific rattlesnake, a red diamondback, a California mountain kingsnake, a California kingsnake, or a two-striped gartersnake—all as characteristic of Southern California as a view of the Hollywood sign on the Santa Monica Mountains.

Back east, have a Primanti Bros sandwich—an incredible combination of beef strips, coleslaw, and fries, all piled between two buns—and then go out looking for a short-headed gartersnake, copperbelly watersnake, massasauga, foxsnake, or Kirtland's snake. I recommend looking in early October, when the nearby ridge will be a conflagration of fall colors. While you're looking for those rarer snakes, you're certain to find an eastern gartersnake, northern watersnake, queensnake, ratsnake, timber rattler, or ribbonsnake. These snakes are as characteristic of the glacially sculpted eastern forests and meadows as the Rustbelt towns, pastures, and ballparks surrounding them—as American as Hoosier basketball, Buckeye football, and the Pittsburgh Pirates.

Down south, those meat-and-threes—places that serve slabs of meatloaf or fried steak next to your choice of three vegetables, all ruthlessly drowned in butter—are never too far away from some dark wetland ornamented with Spanish moss, a swamp that will have a list of truly American snakes as long as your arm: yellowbelly, northern, banded, green, and brown watersnakes, black swampsnakes, glossy swampsnakes, and queensnakes—all remarkably part of the same evolutionary assemblage. These share the swamp with ribbonsnakes, gartersnakes, timber rattlers, pygmy rattlers, and cottonmouths. Atop the nearby sand ridge, with its carpet of wiregrass and stately longleaf pine—down by the broken-down rusty shack and 1982 Trans Am up on blocks, never too far from the aroma of chicken parts frying in Crisco—if you're good, you'll find an entirely different set of spectacularly American snakes: pinesnakes, scarletsnakes, scarlet kingsnakes, chain kings, and surely a white oak runner. Perhaps there too you'll find the rattler that should be the pride of Southerners and the envy of the world, the largest rattler of them all: the dignified and adamant eastern diamondback.

Or you might find yourself dipping some fries in a tangy mayonnaise sauce at a burger joint in Mormon country, taking in the spectacular vistas of wind-sculpted rock and gnarled pinyons after a weeklong float trip, and if you do, be sure to poke around a bit and you might discover a midget faded rattlesnake, Grand Canyon rattlesnake, gophersnake, black-necked gartersnake, wandering gartersnake, or striped whipsnake—snakes as characteristic of our great western national parks as their sandstone arches, trackless canyons, and ice-sculpted peaks.

These snakes are as American as apple pie.

2 · Form and Function

The Way a Sidewinder Moves

To know how a sidewinder really moves, you must first visit the bleakest nightmare of a sandblasted, fiery desert. I drove my field class across the Mojave National Preserve up to Baker, California, home to the world's largest thermometer, which that afternoon read 116°F. We descended from an elevation of several thousand feet down to just around 800, and watched as the vegetation dwindled from relatively lush Joshua tree scrublands to creosote flats and finally to damn near nothing at all. Geographers from other continents scoff at the American notion of what constitutes a desert, and dutifully classify our barren Great Basin as a semidesert shrub steppe instead. They call the giant saguaro forests of Arizona a thorny succulent savannah. But even these cynical foreigners shudder in awe of the raw sterility of Death Valley and its environs. The Mojave is our most intimidating desert, and a pall seemed to come over the students when we arrived there. They would all later agree this

The iconic sidewinder of the Desert Southwest. *Photograph by Marisa Ishimatsu*

part of our trip was the low point. Steinbeck wrote, "The Mojave is a big desert and a frightening one. It's as though nature tested a man for endurance and constancy to prove whether he was good enough to get to California."

Baker materialized, tucked between two giant salt flats. It was a majestic if brutal landscape: jagged brown desert mountains, crags of old black lava fields, shadows folding into tawny salt flats. To the left was Soda Lake and on the right Silver Lake, within valueless Bureau of Land Management property. Some of the students heard me talking about Soda Lake and anticipated we would perhaps go swimming in a nice California reservoir named for sweet carbonated beverages. Not so. Instead: blasted, barren, lifeless salt pans for miles. Silver Lake too is entirely evaporated, its flats devoid of life and ringed by only the hardiest desert survivors.

We stepped out of the vans, and my eyeballs immediately became wicked dry by the pounding hot wind sweeping across the flats. My inner nostrils quickly acquired the texture of corn chips. We walked along the shores of the sepulchral Silver Lake and identified just four shrubs: two kinds of saltbushes; white bursage; and the venerable, scraggly, green-brown creosote, which barely penetrated the edge of the salt flats. This hardy shrub, which covers millions of square miles of our most inhospitably dry desert terrain, was much more common about 1,000 feet upslope, where conditions were presumably less hostile to plant life. There is a well-known effect of elevation on plant life in the Southwest; the higher you go, the wetter the environment. The peculiar thing about the California deserts is how you must travel a few thousand feet upslope to encounter any desert vegetation; at low elevations, nothing grows. At elevations as high as 4,000 feet, simple shrubs and cottontop cactus are all you can expect.

The white encrusted lake floor stretched flat forever, and the sun slowly flattened orange behind the base of tilted mountains the color of sulfur. We drove back to camp at dusk and stopped for a potential creature spotted on the road—what seemed to me perhaps a gecko or another lizard but turned out to be the twisted bean pod of a desert wash plant. I waited impatiently to see what the students would report, guessing that whatever critter we'd stopped for was probably now long gone.

Then I heard a student shout "rattlesnake!"

I grabbed my snake tongs from under a pile of field guides and pamphlets and sprinted back to the other van. I got there just in time to see the pale form of a small snake escaping under the base of a creosote. We stopped for the bean pod, but there was a snake on the shoulder we hadn't seen. It was perhaps a foot long, pale tan with darker tan blotches narrowing to still darker tail bands. Unmistakably, a sidewinder.

Sidewinders are the classic American snake of our most barren deserts. Found only in the lowest, hottest flats of the Sonoran and Mojave Deserts, this hardy little snake spends most of its time curled up half-buried in sand under the meager shade of small shrubs. At night, they spring into action, moving considerable distances over dunes and shrub flats seeking good ambush sites. By day they ambush lizards, and by night they ambush small rodents. Their venom is mild for a rattlesnake—if you're going to be bitten by a rattler, this is the one to get bit by.

I captured it and coaxed it safely in a clear plastic tube. The students admired the handsome face and "horns" of this true desert dweller—sidewinders have enlarged, triangular scales above the eyes whose function is still a mystery—and as is often the case when pitvipers are restrained, the students were able to see the fangs as the snake struggled to defend itself in the tube. The fangs looked like tiny glass shards. All nine students—including young Alejandra, who was deathly afraid of snakes—safely touched the tail of a live rattlesnake.

Sometimes you just need to see something for yourself before you can gauge how a creature moves. After I released the sidewinder, we staged a few photos in the more natural setting of a creosote gravel flat. That's when the snake surprised me. It began unfurling into its indescribable, sideways eponymous movements and deftly crossed about 4 feet of desert floor in a few seconds. It moved like a windup toy; a windblown feather boa; a temporary light reflection on the ripples of a moonlit lakeshore; water traces carved by a paddle on the river. I thought I knew how a sidewinder moves, but I hadn't the remotest clue how fast or uniquely this snake glides across the desert. It could even jump—at one point on the curl of a single coiling repeat, the snake lunged forward and skipped about a body length ahead—like light glancing across water, skipping waves.

We now understand step by step the biomechanics and the exact pivot points and flexions that these snakes use to move across smooth sand. We know from a bioenergetic standpoint that there is no more efficient way to move across such an uneven and giving surface. I had seen movies depicting sidewinders moving across sand, and I had studied the diagrams of these movements and how they relate to the tracks left behind. But nothing in my experience could have prepared me for seeing firsthand how quickly and effortlessly a Mojave sidewinder can make its escape into the desert night.

The way a sidewinder moves defies description but owes everything to the limbless form of the snake's body and to its formidable environment. Snakes are the only animals known to move by this unique locomotion, whose smooth and elegant strides leave unique tracks in the sand and allow rapid overland movement across vast stretches of open desert. I can think of no better example of how form and function converge to produce such a simultaneously graceful and effective adaptation, which enables existence within one of the world's harshest environments.

This chapter focuses on the unique body of snakes. At first glance they may appear strange and alien compared to more familiar animals, yet closer inspection reveals an unexpected kinship with humans. The rest of the chapter deals with how this unique body performs in its environment, including the cold-blooded physiology of snakes and the many ways that snakes move.

What better way to tour the body of a snake than beginning at the snout and venturing down its long serpentine course to the tip of the tail? First I begin by pointing out the rather unobvious fact that snakes do indeed have a head, body, and tail. This is not an animal with a head attached to a three-foot tail. The tails of most snakes are actually pretty short. Most of the length of a snake is body, followed by a tail that makes up perhaps only 10% of the total length.

The Business End

Common sense dictates that for an animal with no limbs, it is important to place all sensory structures toward the snout. Snakes cannot feel their way through the dark or shield their face with hands or whiskers, so there is a considerable devotion to the sensory structures of the head. The head of most snakes is made up of large plate-like scales that for many species serve as keys to their identification. Some rattlesnakes have reduced head shields, and large areas of the head are instead covered by small body scales, which give their head a unique appearance. Several of our southwestern desert species have uniquely enlarged snout scales and jaws that sink snugly within the snout, both of which are apparent adaptations for digging face-first through sand. Many snakes have unique numbers and arrangements of scales with rather arcane-sounding names that refer to regions of the head: occipitals, loreals, postoculars, labials, temporals, nasals, infralabials, and so on. If you have no clue what any of these refer to, you're in good company; I've never bothered to memorize them all.

The nasal scales of course refer to the scales that border the nostrils, which in snakes work much the same way yours do. Most of the time snakes breathe through their nostrils and not through their mouths, and the nostrils lead to a small and moist cavity where airborne chemicals can be detected—this is the same sense of smell that you have, and the same olfactory bulb of the brain is responsible for smelling whether you are a snake, an aardvark, a largemouth bass, or a trial lawyer.

When flicking the tongue, a snake's second chemical sensory modes reach far forward beyond the snout, often farther than the entire length of the head. The tongue is one of the snake's most recognizable and misunderstood attributes. Contrary to myth, the tongue cannot inflict harm, nor can it hypnotize. However, I confess that its movements can be hypnotic. The tines of the tongue can be spread wide and waved forth, or held in place for a moment in an almost threatening gesture. The actions of the tongue differ given the situation: fast, rapid, rhythmic pumping while searching for a recently bitten mouse; slow, steady flicking when hunting for fish; long, static strokes when threatened. Snake tongues can even differ between the sexes: male copperheads have longer tongues of a different color compared to females.

An eastern hog-nosed snake flicks its substantial tongue. *Photograph by Bob Ferguson*

The snake's tongue works slightly differently than that of humans. Rather than being covered with taste buds, it picks up nonairborne chemicals from the environment and then quickly retracts into the mouth. Snakes have a large and active vomeronasal organ—a special bulb on the roof of the mouth that connects to the olfactory bulb. When retracted into the mouth, the tongue lies flat and an airtight seal surrounds it. The shape of the oral cavity is the perfect companion for the tongue at rest. With a quick squeeze shut, the fluid around the tongue is forced rapidly up a duct to the vomeronasal organ. This occurs every time the snake flicks its tongue and retracts it back into the mouth.

Vomeronasal organs are also found in mammals, and they are well studied in laboratory mice and other rodents. It is still something of a mystery as to whether humans have one that works. But consider this: the key to rodent sexual behavior and attraction (and even monogamy in a few species) has much to do with vomeronasal organ function. This kind of chemical communication is referred to broadly as pheromone communication. Pheromone communication in humans is still controversial, but there is some evidence that we do respond to chemical signals from each other, albeit subconsciously. Have you ever noticed how strongly your emotional attraction or memory of a loved one can be triggered by a good, hearty whiff of a scent you associate with that person? This also appears to be the case with snakes. The vomeronasal sense in snakes is associated with pheromone communication, attraction between males and females, and perhaps also adults and youngsters. It also plays a key role in establishing hunting grounds, tracking prey, and recognizing familiar surroundings.

With one-third of their brain and two entirely different organs devoted to sensing their environment's chemistry, the sensory world of snakes is a truly strange one when compared to our meager chemosensory abilities. It is difficult to imagine what seeing the world through smell would be like. But most people who have owned dogs have probably tried to imagine this. Snakes and dogs share an excellent sense of taste and smell, and their capabilities are probably fairly comparable.

Traveling down the snout toward the eye, some snakes have a bizarre sensory structure not known in most mammals. But the heat-sensing pit characteristic of pitvipers is not entirely unprecedented. The actual sensory neurons found in this organ are not unique. You have them in your skin, and they are especially well represented on your hands and face, and even your lips and tongue. Snakes have these temperature-sensitive neurons as well. But in some, if not most, snakes, they are concentrated on the face because they experience the world face-first. A few groups of snakes have developed large concentrations of heat-sensitive neurons in organs that maximize heat sensation. American rattlesnakes, copperheads, and cottonmouths are adorned with the most efficient heat sensors of them all.

Pitvipers should receive praise for their heat-sensing pit, which can be considered one of the great superlatives of vertebrates, equivalent to the eyesight of hawks, the sonar ability of whales, the echolocation of

The heat-sensitive pits of this pygmy rattlesnake are noticeable on either side of the tongue.
Photograph by Matthew Sullivan

bats, and the flight of a hummingbird. The sophistication of this organ cannot be overstated. It is an extremely thin membrane suspended in a pit on the face between the nostril and eye. This membrane has a pocket of air behind it, with its front in contact with the outside world. The membrane is served by a remarkably complex and intricate capillary bed, which, when viewed in thin-section microscopically, resembles a complex spider web. The air behind the membrane retains the approximate body temperature of the snake, and abundant nerves within the capillary bed detect tiny shifts of air temperature in front of the membrane. The pit is served by the trigeminal nerve—the same one inside your face—which transmits this temperature information to the optic tectum in the brain. This is where the real magic happens: the thermal arithmetic detected by the pit is computed to the nearest hundredth of a degree. Finally, input from the eyes and the facial pit is combined to form an overlapping visual/thermal image, *exactly* like the alien from the Schwarzenegger movie *Predator*.

Continuing below from the pit you might expect to find ears, but in snakes they're missing. Among the ways to distinguish legless lizards from snakes is the presence of external ear openings. Of course, even in lizards these are not the fine, expressive, furry ears of mammals, but if you look carefully, you will find a hole on the side of a lizard's head. Lizards can hear airborne sounds as well as faint vibrations humming through the soil, at the approximate frequency of a human footfall. Snakes do not have external ear openings, although they do have a capsule at the same location of the inner ear of lizards. This contains the semicircular canals and vestibule, which allows snakes to maintain their sense of balance—a necessary feature for the limber acrobatics of tree-climbing snakes. Their inner ear also enables snakes to feel ground and perhaps airborne vibrations, such as the approach of predators.

The eyes of snakes are usually held in sockets on the sides of the head, allowing snakes to monitor potential enemies coming from any angle. But some day-active snakes have their eyes in a more forward position so that they have nearly binocular vision—eyes flat on the face in front allow an overlapping image complete with depth perception. These are snakes such as coachwhips, whipsnakes, and racers, which will typically meet your gaze if you confront them, and may even strike at your face in self-defense. The eyes of snakes have no lids and are instead protected by a transparent scale. When the snake sheds its skin, all the scales are cast as a single unit, including the eye scales.

The eyes of snakes are unique, having been derived evolutionarily from a blind ancestor and essentially reinvented (see chap. 1). And it should be mentioned that the immediate ancestors of snakes—lizards—have excellent vision that is better than that of any mammal, including humans. Lizards can see in the ultraviolet part of the spectrum, which humans can only imagine. So it is unusual that snakes lack such an excellent apparatus. Parts of the snake eye develop from different embryonic tissue than other vertebrates,

The forward-facing eyes of coachwhips give them excellent daytime vision with excellent depth perception. *Photograph by Noah Fields*

Nocturnal snakes often have vertically elliptical pupils, like this nightsnake. *Photograph by Curtis Callaway*

the lens is focused differently than in other reptiles, and the part of the brain responsible for orchestrating an image has undergone obvious reorganization. The eyes of snakes are not supposed to be especially keen, and many species lack the cones typical of other vertebrates. Cones are special light-sensitive neurons that require good light saturation and distinguish color. Rods are also light-sensitive but work well in dimmer light and only distinguish grayscale. The cones of many snakes differ from those of other reptiles, and some researchers have assumed snakes cannot see in color. But some day-active snakes do have modified cone-like rods with color pigments that may support rudimentary color vision. And many nocturnal snakes with big eyes have rod densities exceeding those of humans and comparable to nocturnal mammals such as opossums and cats, which gives them excellent night vision.

Snake senses are processed by a rather small brain, whose proportions suggest that snakes experience the world mostly from the perspective of its chemistry. They smell and taste their world through their nose and vomeronasal sense. Vision and heat are secondary yet well developed in some species. But the rest of the brain is still there. You may have heard the outdated concept that humans have a "reptilian" brainstem to which a large and complex "mammalian" cortex has been added, like a huge scoop of ice cream added to an undersized cone. This view has been replaced by the much more accurate and interesting notion that all vertebrates—from catfish to plumbers—have the same basic brain segments and hardware. The sections of the brain—divided most simply into forebrain, midbrain, and hindbrain—share similar functions and mechanics in all vertebrates. From this basic, successful plan, inherited from the earliest vertebrates, some innovations and elaborations arose. Humans have a gigantic forebrain cortex useful for sensing the world through the hands and face and for language, massive memory storage, and development of abstract notions. As special as our cortex is, all vertebrates, including snakes, share a subtler version. For their part, snakes have an impressively enlarged olfactory bulb, and some have an impressive optic tectum. Compared to humans, the olfactory bulbs of snakes are massive.

The brain of snakes is enclosed within a hard and compact braincase, another holdover from their subterranean ancestry. Many burrowing reptiles have similarly hardened skulls, better for battering their way through compact soil. I experienced firsthand the well-guarded brain of snakes when I dissected several for a research project. Exposing the brain took careful excavation using a powerful but precise snipping tool, and there was no easy point of entry.

Oral Weaponry

Contrasting starkly with the heavy-duty braincase of snakes are the delicate and flexible jaws and teeth attached to it. They are among the most unique attributes of snakes. Both the upper and lower jaws of snakes are extremely flexible and swing wide along hinges connected by ligaments. The most evolutionarily advanced snakes have the most elaborate and flexible jaws, which allow them to move their faces with unparalleled mobility. It's almost as if they have a pair of arms slung under their heads, which can reach out and around their food, pulling it down their throats. The upper and lower jaws work just like that, actually reaching forth and pulling back in an alternating fashion, like a swimmer's breaststrokes. The skin between the jaws can stretch incredibly, and a tubular trachea extends like a snorkel to the tip of the lower jaw, allowing intake of air while the food is being worked down.

The jaws of most of our snakes are equipped with several rows of small, exquisitely sharp, recurved teeth. A bite from these snakes is harmless and

Harry Greene

Getting in the Snake's Skin

Harry Greene takes his time, publishing unconventional papers that ultimately become classics. His reputation has developed from his keen insight, his ability to eloquently describe patterns, and especially his cult of personality. When Harry gives talks at conferences, the room is packed with people standing along the walls and squatting in the center row. When you have dinner with him, the whole table is transfixed by his stories and laughs at his jokes. There is just something about Harry. He's a great guy.

Photograph by the author

Harry got his start out on the Texas prairies where he caught snakes as a boy. He published several papers in scientific journals when he was still in high school. After a stint in the army as a medic, he went back to school and began publishing some of the most noteworthy papers on snakes anyone has seen. He studied bushmasters in Costa Rica, black-tailed rattlesnakes in Arizona, and timber rattlesnakes in New York. But his book about snakes propelled him to rock-star status. In my opinion, *Snakes: The Evolution of Mystery in Nature* is the only book written by a herpetologist that could rightfully be considered literature. It was a massive hit with the public and scientists alike, and I highly recommend it. But first sit back down and finish reading my book.

I think Harry's biggest contribution to snake biology is his research style. I remember him talking about black-tailed rattlesnakes, about how he wanted to get down on his belly and understand the snakes from their point of view. He's not only interested in the lives of snakes, but also what goes on in their minds. I read Harry's book at the outset of my career, and have always tried emulating his attempts to push the envelope (and to a large extent I failed). Harry takes his time and puzzles over his work, dreams about it, seemingly more interested in the *feel* of a study than the results. His work has a real emotional component, as revealed in his latest book, *Tracks and Shadows: Field Biology as Art.*

One afternoon, I had the pleasure of walking out in the oaky hills near Ithaca to see Harry's timber rattlesnakes. We met with the landowner, who was a big Upstate New Yorker wearing Wellingtons and a flannel shirt. Harry talked with him, and with his white beard and easy demeanor could easily have been the man's uncle. The only evidence that he was instead an Ivy League professor was Harry's bright red Cornell cap. It was a violet spring day, sunny and warm, but the trees were still bare and clouds purple-centered and ringed silver. As we approached the rock outcrop, Harry spotted the snakes first, pointing out a flattened yellow timber rattlesnake sandwiched between two Devonian slabs of sandstone. Like all pitvipers, it simply sat there, soaked up the early sun, and ignored us. Harry took a few pictures from his belly, the stance he's often advocated.

feels like getting scratched by briers or a rosebush. The bite also reveals the number and arrangement of their teeth: a good one will break the skin and cause painless droplets of blood to appear. These wounds correspond to rows of teeth on the maxilla, mandible, and roof of the mouth. Recurved teeth help the snake hold on to slippery prey and assist in guiding the meal toward the back of the mouth to be swallowed. They also make it difficult to extract snakes from your hand when they bite, and care must be taken to pry an ornery snake from your skin and simultaneously avoid damaging the snake's jaws.

In addition to the small teeth shared by all snakes, there are three main types of fangs possessed by venomous species. The simplest fangs are the enlarged, grooved teeth in the rear of the mouth of some species within the harmless snake families. These are technically venomous snakes, although it should always be recognized that they are harmless to humans. Dipsadines such as hog-nosed snakes, ringneck snakes, mudsnakes, and wormsnakes, as well as colubrids like crowned snakes, have these opisthoglyphous, or "rear fangs," and some conduct mild venom from a gland that assists in the subjugation and digestion of slippery prey such as lizards, toads, salamanders, centipedes, and worms. Some of our poorly known southwestern snakes have rear fangs and mild venom and should be treated with measured respect. The lyresnakes, somewhat large and mysterious desert dwellers whose distribution barely enters the southwestern United States from Mexico, have bright bug eyes, elliptical pupils, and a rounded head separated from the skinny body by a distinct neck. If it weren't for their extremely lanky proportions, they might be mistaken for a pitviper. And the venom can cause alarming symptoms—a friend of mine recently captured one that dutifully rewarded her good luck with a nasty bite on the knuckle that left her swollen with shooting pain that lasted until the following morning.

The skull of a gartersnake, showing the delicate jawbones and teeth. *Photograph by Noah Fields*

Elapids such as coralsnakes and seasnakes have a more efficient venom delivery system composed of a pair of enlarged, hollow fangs in the front of the mouth. These are served by a duct leading to a gland in the upper jaw that produces potent venom. The venom is high in quality but low in quantity, and the front-fanged proteroglyphs are often assumed to require a period of chewing before dangerous quantities of venom can be injected. Compared to pitvipers, this may well be true. But never assume that a coralsnake cannot deliver a nasty bite. They seize lizards and small snakes and quickly incapacitate them with the venom. For the seasnake, a quick bite to a dangerous and spiny fish is sufficient to render them helpless for swallowing.

The most efficient venom delivery system is the hinged, hollow fang of vipers, the so-called solenoglyphous fang condition. Our pitvipers—including all rattlesnakes, the copperhead, and cottonmouth—share this state-of-the-art weaponry. At rest, the enlarged fangs are folded flat on the roof of the mouth pointing toward the throat, with most of the length wrapped snug in pink gums like some hideous pig in a blanket. Copperheads and cottonmouths have fangs about as long as a cat's claw. Eastern diamondback rattlesnakes have the largest fangs of any American snake—about a half-inch long and curved into a cruel yet elegant scythe. When wielded, the entire front section of the upper jaw swings forward owing to the lever action of a rear jaw bone, and the fangs are erected perpendicular to the mouth, not unlike a switchblade. The fang can rapidly penetrate the body of prey or foe, and its sharp tip and hollow central lumen allow copious amounts of venom to be squirted deep into tissues, in precisely the way a hypodermic needle administers vaccines or drugs.

The venoms of snakes with either of the three fang conditions are a complex blend of proteins, enzymes, and other factors whose primary function is rapid immobilization and digestion of prey, but whose secondary function —defense—has captured the imagination and spread fear in the hearts of humans for generations. Chapter 9 is devoted to explaining the significance of dangerous snakes and their venoms.

The Sinuous Body

We now leave the head of snakes, which is by far the most interesting and complicated part of their body, and continue our tour down and within the body's coils. Here snakes show interesting novelties when compared side by side with standard-issue vertebrate features. The outward appearance of a snake's body is immediately recognized by most people owing to its long, tubular shape thrown into coils, which are most apparent when the snake moves. The shape and characteristic movements of snakes are produced by repeating vertebrae and ribs connected to an impressive array of rectangular deep muscles and sheets of shallower trunk muscles running the entire length of the body. The repeating lateral and abdominal muscles move in concert and provide the snake with its seamless, rhythmic motions.

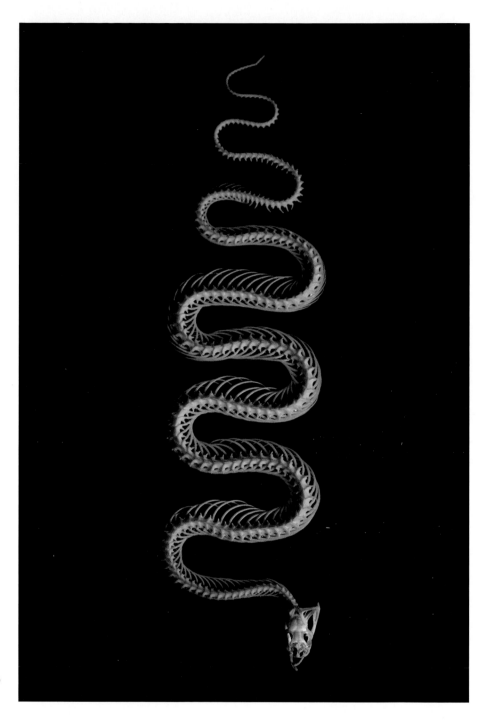

The skeleton of a gartersnake. *Photograph by Noah Fields*

Overlaying these muscles is loose skin covered with body scales of similar size and shape. The scales are typically oval and overlap to form a tough, waterproof covering. The skin can stretch between the scales, allowing accommodation for large meals or heavy breathing. On the belly, most snakes have a long series of rectangular ventral scales that differ in shape from the body scales. The snake's skin is shed as a complete unit as the snake grows, often as many as three or four times a year.

The body scales differ slightly in number, size, shape, and texture among species and can aid in the snake's identification. One of the most obvious differences among snakes is whether the scales are smooth or have a raised ridge, called a keel. Snakes with keeled scales have a rough appearance, while smooth-scaled snakes often appear shiny and have the texture of a nice leather couch. Glossy snakes, for example, are so named because they have smooth scales. It is difficult to describe the appearance of a glossy snake, which otherwise looks much like other blotched terrestrial snakes of the American grasslands and deserts. But in hand they are instantly recognizable by virtue of their silky-smooth texture, which never fails to provoke vocal admiration from my students.

Snakes have a unique feel. This is not something you can quantify or refer to objectively, but snakes feel good in your hands. It's as simple as that. They are not slimy. They are muscular, and the scales are often smooth and nice—the cool, smooth-leather interior of a sports car is the closest analogy I can provide—and they feel supple, powerful, and wonderful. Tame snakes seem to like the body heat in your hands, so they enjoy being handled and gently cling to your hands, arms, or neck while moving around. I've witnessed first-hand the excited chills caused by their gentle wanderings in novice as well as experienced snake handlers. The responses have sometimes been downright erotic. Do yourself a favor. If you've never touched or held a snake, find somebody who has one, go to a pet store, or see a live snake demonstration at your nearest zoo or aquarium. You won't regret it.

Under the remarkable skin, the body of a snake is simplified into an elongate trunk without a distinct chest and abdominal cavity. The heart is contained within a membranous sack, but all the ribs and vertebrae of snakes are essentially the same size and shape, start behind the head, and continue uninterrupted to the beginning of the tail. This is somewhat different from the situation in mammals, which have a cavity where the ribs enclose the lungs and heart and are separated from a ribless abdominal cavity that holds the guts. Besides this minor difference associated with the need to streamline movements of the coils, the innards of snakes will be familiar to anyone who has gutted a fish, cleaned a deer, or spent any time in a freshman biology class.

The digestive tract is a simple, elongate tube beginning at the mouth, continuing into a powerful stomach below the heart, coiling modestly into a simple intestine, and ending above the tail. It is associated with familiar digestive glands—a great crimson liver, bulbous green gall bladder, and pancreas spread through the intestinal membranes like old, heavily chewed bubble gum. These glands secrete enzymes used in breaking down the proteins, fats, and other components of the snake's meal. But these familiar features—shared by most vertebrates—belie an incredible process. Because many snakes eat infrequent yet large meals, their digestive tract is usually in a state of torpor: degenerate and idle. When they do feed—sometimes consuming prey as large as or even

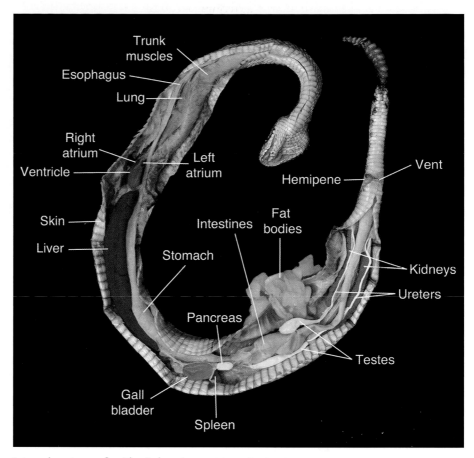

Internal anatomy of a sidewinder. *Photograph by Noah Fields; digitization by Crystal Kelehear*

larger than themselves—the digestive system kicks into overdrive. The mass of the organs increases by several times. The digestive enzymes produced by liver, pancreas, and the gut itself can increase by several orders of magnitude. For years, digestive physiologists found it difficult to understand the processes involved in chemical digestion because they studied animals that feed frequently upon small bits of food—animals like humans and rodents. When these animals feed, the stomach, glands, and intestines barely register a difference. When digestive physiologists began to study pythons, it led to a revolution in our understanding of how the process works, because in these and other snakes the events are extreme and explicit.

A normal meal can be processed completely and eliminated in as few as three days in ratsnakes to as long as two or three weeks in pitvipers like sidewinders and timber rattlesnakes. Typical snake scats are brown, with large amounts of white urates, and contain hair if the prey was mammalian. Ratsnakes are able to excrete the shells of bird eggs as thin, tapered scats. But most everything else in the meal—meat, tendons, bone, scales, and claws—is digested with high efficiency and devoted to growth, maintenance, and reproduction in the snake.

The heart of snakes is fairly large and centered about one-eighth of the way down the body from the head, depending on the species. You can feel a snake's heart by gently squeezing the snake like a tube of toothpaste between two fingers; traveling down from the head, it is the first solid lump you'll feel, and you can feel it beating. The heart has only three chambers: two atria and a single ventricle. The right atrium receives blood from the veins and quickly passes it to the ventricle. The left atrium receives blood from the lungs before passing it to the ventricle. The ventricle is the largest chamber and simultaneously pumps blood either to the lungs or to the rest of the body. The ventricle is not completely partitioned into two chambers like ours because it doesn't have to be. Only the high metabolic demands of warm-blooded animals requires separation of blood into entirely separate oxygenated and deoxygenated pools. In snakes, the oxygen-rich blood returning from the lungs and the deoxygenated blood coming from the veins can mix in the ventricle. It contains pockets that help prevent total mixing, and in active snakes, pressures help conduct oxygen-rich blood to the tissues and deoxygenated blood to the lungs. So there is no real drawback to having a single ventricle for animals with a slower metabolism. But there is actually a benefit to the arrangement. When hibernating, basking, or diving, snakes can shift all blood away from the lungs toward the outer extremities of the skin and muscles. This, coupled with their low metabolism, allows rapid warming of the body while basking, and efficient use of minimal oxygen while underwater or during seasonal sleeps.

Next to the heart is the first section of a single lung. This is a unique arrangement associated with the tubular plan of the snake body. Primitive snakes like blindsnakes have paired lungs similar to most air-breathing vertebrates. Advanced snakes, including most American species, have only a single right lung that has stretched to fill most of the elongate body cavity. Boids show an intermediate condition; their right lung is enlarged, and the left is rudimentary and practically useless. The upper section of the lung near the heart is red and infused with blood vessels, and it also contains small pockets to increase surface area; this section has the appearance of bloody Swiss cheese and is the only part responsible for respiration. Curiously, the lung continues down the body to fill much of the space not occupied by the other organs; here it lies alongside the liver, guts, and all the way down to the reproductive organs. This sacular portion of the lung is transparent and lacks blood vessels, and it does not exchange gasses. The sack lung of a cottonmouth can be inflated and permits buoyancy, like a built-in inner tube. Pinesnakes can inflate their lungs and then expel the air to produce an impressive hissing sound. The glottis has a thin, fleshy knob in front of it, so the air vibrates around it like a musical instrument. When captured or approached, many snakes will blow themselves up like a balloon to make themselves seem bigger than they are, or to make themselves harder to swallow.

The final occupants of the snake body cavity, which first appear about two-thirds of the way down, are the kidneys and reproductive organs. In snakes they are paired, intimately associated, and served by the same ducts. The gonads are always found in front of the kidneys, so their products are produced and move back toward the kidneys and the rear of the snake. The kidneys and gonads are shifted relative to one another, the right kidney-gonad tract lying several centimeters forward relative to the left. This is another modification not seen in other vertebrates that is associated with the thin, tubular shape of snakes. More will be said of snake reproduction in chapter 5.

The intestines, kidneys, and gonads all produce products (feces, uric acid, and either eggs or sperm) that are excreted by the snake from a single opening near the tail. The cloaca ensures that when you handle a snake it can excrete a disgusting brownish-yellow paste that is part feces and part uric acid, the whitish crystalline material the kidneys of reptiles produce. This nitrogenous waste product is slightly difficult to produce, but it is nontoxic and doesn't require water for packaging, making it a more water-efficient waste product than mammalian urine. The cloaca is a muscular cavity covered by a single or paired enlarged scale referred to as the vent. If you handle a snake, you can turn it over and quickly determine where the tail starts because the enlarged vent scales indicate the end of the line for the internal organs. Below the vent is the beginning of the tail.

An Interesting Tail

Our tour of the snake doesn't end at the vent. The tails of snakes contain some surprises, and some species have developed interesting modifications. Active and maneuverable snakes have correspondingly long and skinny tails. Coachwhips are a classic example, their name being derived from the tethered, bullwhip appearance of their long, skinny tail, which often contrasts in texture and color with the rest of the body. Rough greensnakes are probably our most arboreal snakes, and their long tail assists their fluid movements through the canopy of trees and shrubs. The yellow-bellied seasnake's tail is flattened along the side and shaped like an oar for graceful swimming in the open ocean. Some snakes, such as ringneck snakes, will coil their tails and present brightly colored undertail markings to predators as a distraction: the tail of snakes is short and expendable, and many snakes use it in an attempt to draw a predator's attention away from the head. The tails of rubber boas are stumpy affairs often covered by scars. When threatened, they hide their head among knotted body coils and thrash their blunt tail in the air as a decoy—a second "head."

The tips of snake tails often have a modified scale or series of scales whose function is uncertain. The dipsadines have a conical, spiny tail tip that they often harmlessly dig into the hands of their human admirers. The probing of a large adult mudsnake can be disconcerting but never causes more than a slight pressure. Yet this seems to have been the source of their legendary reputation as "stinging snakes." The actual function, if any, is unknown, but some speculate that it may allow them to hold on to their slippery prey. Or perhaps it helps them cling to underwater structures while their bodies sway in the current waiting for prey. Other nonvenomous snakes have tails that end in enlarged, flattened scales that may aid in sound production when they vibrate their tails in self-defense. And of course the final scales of a rattlesnake's tail are retained and become hardened after each shed, producing the interlocking hollow structures that make up the buzzing rattle.

Within the base of male snakes lies a special surprise. Those unfamiliar are often astounded that you can tell the difference between a male and female snake at all. But the base of the tail is noticeably thicker in male snakes relative to females, and the sexes can be differentiated with practice. The reason for the thicker tales in males is an even bigger surprise. Tucked inside-out in the meat of the tail is the male reproductive organ called the hemipenes. There will be more on this spectacular oddity later (chap. 10), but for now suffice it to say that when fully unfurled, the hemipenes are a truly ugly, fantastically hideous, four-headed spiny monstrosity.

An Efficient Metabolism

That snakes are cold blooded is a misconception too commonplace for me to attempt to overturn it. But it is misleading to call snakes cold blooded, and the association between cold bloodedness, ruthlessness, and bloodthirstiness adds to the already bad reputation of these retiring creatures. Some examples may better make the point. Consider a black racer out in the open and warm to the touch during a cool December day. Or a coachwhip in the Sonoran Desert with prolonged, high body temperatures that would render a human being brain dead. These are not cold-blooded creatures. Still, the different lifestyles of reptiles and mammals are largely attributable to their metabolic rates and how they acquire and maintain body heat, so understanding exactly how they differ is important for understanding how snakes work. Biologists use a different term when referring to the metabolism of snakes and other reptiles —ectothermy—which is more fitting and accurate.

The term ectothermy means "heat from outside," referring to the fact that snakes gain heat from their environment. Biologists refer to mammals and birds that maintain fairly constant and high body temperatures as endotherms, meaning "heat from within." Endothermy requires huge amounts of fuel and oxygen, so endotherms must breathe frequently and have a rapid heart rate to efficiently move the oxygen around the body. Snakes have resting metabolic rates, heart rates, and breathing rates a fraction of that of mammals. Instead, to have a high body temperature, they must get it from their surroundings. In snakes, crucial functions such as metabolic rate and heart rate respond reflexively to outside temperature; a gophersnake's heart beats fewer than 20 times

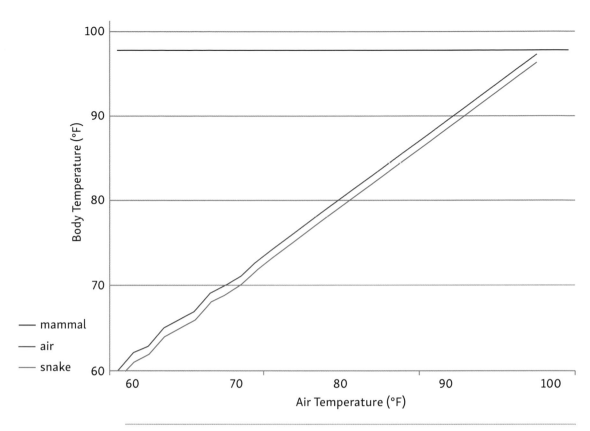

Endotherms and ectotherms have different responses to environmental temperature.

per minute if held at 68°F, but will increase to 60 beats per minute when held at 95°F. At very low temperatures the heart rate of snakes plummets to only a few beats a minute or less, and they enter a state of suspended animation, or torpor.

This is not to say that snakes always have variable temperatures or cannot maintain constant body temperatures. In fact, some snakes are remarkably good at maintaining constantly high body temperatures. But instead of using physiological methods to keep their body temperatures steady, snakes use behavior. Like endotherms, snakes have a thermostat in their brain and a set point that corresponds to their preferred body temperature. But when they begin cooling off and their blood temperature falls, they simply bask in the sun to quickly make an adjustment. When they heat up too much, they duck into the shade to cool off.

It is natural to assume that because humans are endotherms, our system is the best. For a long time, even scientists believed this, and "cold blooded" was used synonymously with terms like "lower vertebrates" and "primitive" to describe amphibians and reptiles. But if endothermy is so wonderful, how come endotherms are outnumbered by ectothermic species throughout most of the world's ecosystems? The fact is, there are benefits and drawbacks to both systems. But it is better to think of them as two different, yet equally effective, means to survive.

The benefits and drawbacks of endothermy versus ectothermy largely come down to the amount of energy required to run the two systems. Endothermy is wasteful. Ninety percent of the food you eat is burned up to support your high metabolic rate and high body temperature. Ectotherms much more efficiently convert food into body proteins and fats, and they waste little of the food they eat because they do not require large amounts of calories to support a high body temperature. By contrast, they can maintain a constant and high body temperature simply by moving about their habitat gaining and shedding heat.

For snakes, the drawback of ectothermy is that it can often be difficult to maintain a favorable body temperature using behavior alone. Some habitats may not contain sunny areas that allow certain species to thrive. Other habitats, such as severely cold arctic tundra, preclude snake activity except for perhaps a couple of months a year, and this is clearly too short a time period to support growth, reproduction, and a viable snake population. It also takes some time for snakes to gain or shed heat, when a similarly sized mammalian predator essentially wakes up every morning ready to go.

There is an advantage to maintaining a body temperature in the vicinity of 98.6°F, since most mammals and birds have their thermostats set in this neighborhood, and many snakes also bask until their body temperatures approach the same temperature. Most vertebrates get in trouble at temperatures exceeding 106°F, yet some snakes tolerate normal body temperatures as high as 105°F. Above this temperature extreme, body proteins begin to unravel, and death quickly follows. But warm temperatures of around 100°F are perfect for most biochemical reactions, so digestion, cell growth and division, protein assimilation, and immune function are maximized at such temperatures. Snakes will therefore often bask more frequently under certain circumstances: while digesting a large meal, while gestating developing embryos, and perhaps even when they are sick.

Endothermy and ectothermy also support two different exercise physiologies. Mammals and birds have muscles and metabolic rates that support aerobic exercise, which gives them better endurance. Few ectotherms can achieve aerobic muscle performance, and instead their muscles largely function anaerobically, much like a sprinter. So snakes can often escape rapidly with vigorous movements, but they quickly become exhausted. Exceptions are the coachwhips, racers, and whipsnakes, whose activity levels, metabolic rates, and aerobic performance are several times higher than most other American snakes.

Movement

For elongate, legless creatures the best way to move is by lateral undulations, the wavy side-to-side movements typical of most snakes. This is accomplished by movements coordinated between the numerous muscles attached to the vertebrae and ribs. No fewer than 21 distinctive muscles take part in moving the coils of a snake from side to side. Scientists can tell you in exhausting detail how each insertion of each muscle on each vertebra contributes to movement, and all about the kinetics of each swaying coil. But descriptions of these muscles, movements, and the diagrams that accompany them make my eyes glaze over and have no solid foundation in reality. The truth is, snakes have excellent control of their long bodies. They can throw themselves into sideward arcs that push from the substrate, and these arcs move down the body in waves, creating constant forward motion. The movement is created by the sides of their body rather than their bellies. This they do unthinkingly and with great grace. All American snakes can also swim by using lateral undulations.

Lateral undulation is not unique to snakes. Most elongate animals—from worms, eels, and salamanders to legless and aquatic lizards and crocodilians—use this efficient movement pattern. Most American snakes move by lateral undulations most of the time. They need to push these loops off something solid, so the undulations are used to weave along complicated surfaces such as thick, shrubby vegetation and rocks, and can even allow them to glide through grass. Without substantive obstacles to push from, snakes flail helplessly and can't go anywhere; snakes attempting to move quickly on smooth pavement often wiggle frantically in place.

Rectilinear locomotion is unique to large, heavy-bodied snakes, which use it for slow crawling movements. It is best developed in large, camouflaged snakes that slowly crawl prone along flat surfaces and therefore escape detection despite their size. Rectilinear locomotion involves slight ripples of the

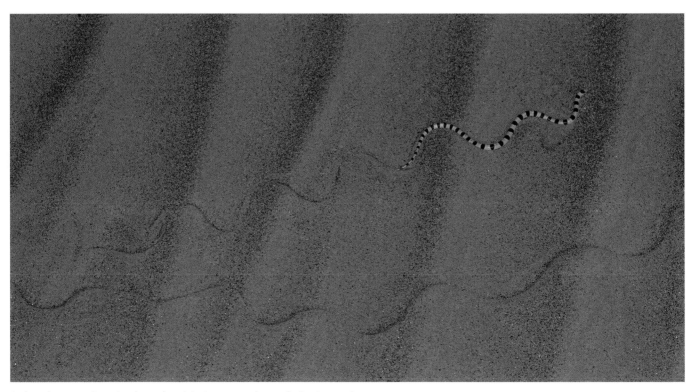

A shovelnose snake using lateral undulations to make tracks in the sand. *Photograph by Zack West*

belly and belly scales, which travel down the body and propel the snake forward like the movements of a millipede. Imagine that instead of having 6-pack abs, your weightlifter buddy had a 136-pack, and he used his many abdominal muscles to crawl slowly along the ground, each pair of abs working independently like a pair of legs. That is how rectilinear locomotion works. Only a few species of American snakes are bulky enough to rely on this form of locomotion. Our native boas will crawl across a desert highway using rectilinear locomotion, and most of our pitvipers occasionally use it as well. Hog-nosed snakes are chubby enough to use this movement pattern, and the southern hog-nosed snake is quite adept at moving slowly across sand this way. This species and large eastern diamondback and timber rattlesnakes habitually use rectilinear locomotion. It is interesting to watch a large snake move this way. From a distance, they appear to glide forth almost like a slug, and it is difficult to perceive any movement at all.

Concertina locomotion is a special movement used by snakes to climb trees or move through burrows. The snake bunches up its posterior coils to brace its head and neck, which reach forward and anchor. The forward section then serves as a pivot point while the rear coils are brought forward. The snake will slink along, and it can make rapid progress. This form of locomotion is common in American snakes that live most of their lives in burrows or that pursue prey in burrows; in tight spaces, snakes reach forward with their head and neck, anchor, then quickly bring up their rear coils. Snakes such as glossy snakes,

A red cornsnake uses concertina locomotion for climbing. *Photograph by Daniel R. Wakefield (www.flickr.com/photos/genesispythons)*

mole kingsnakes, gophersnakes, pinesnakes, and many others use concertina locomotion to explore rodent burrow systems. Tropical arboreal snakes are good at utilizing concertina for climbing, but several American tree-climbing snakes use it as well: greensnakes, brown vinesnakes, ratsnakes, and cornsnakes. For American snakes, a modified, rather clumsy version of concertina is used, usually to scale the vertical surface of a large tree. Ratsnakes are our most accomplished trunk climbers. Ratsnakes also have an additional adaptation that assists their ascent: the cross section of these snakes is not round, as in most terrestrial snakes. Instead it is shaped like a loaf of bread—rounded along the spine and flattened across the belly—and the ratsnake's belly scales can be turned out sideways and wedged into insignificant grooves within the otherwise smooth bark of a tree. In this manner, ratsnakes scale tall trees in a remarkably short period of time, making them the terror of nesting birds throughout the country. Many other snakes are surprisingly accomplished climbers and occasionally venture into shrubs and trees. Even heavy-bodied snakes like timber rattlesnakes and copperheads are occasionally found in the branches of trees.

The fourth and final snake movement is the most specialized, and it allows fluid movement across loose sand. This is the sidewinding mused about at the beginning of this chapter, which various snakes occasionally use to move across uneven, slick, or loose surfaces. For many snakes, these attempts stem from vigorous undulations undertaken when the snake tries to escape. They are not capable of the flawless movements of our desert sidewinder. But sidewinding does appear to be more common among snakes than you might have thought. Copperheads can glide fairly well across the smooth surface of a road, and I've seen cottonmouths rapidly sidewind across greasy floodplain mud. Watersnakes can also sidewind across mud.

Sidewinding involves the snake tilting in its direction of travel and undulating rapidly. The motion is distinct from lateral undulations because the body is actually held off the ground and only firmly touches the substrate at a few key points. In this manner, the coils dance across loose sand almost like the skipping legs of a child, and the full length of the snake's body barely touches the ground. Narrow points of the snake are just skimming the surface, and the movement requires little time or energy. It is by far the most efficient way to travel on sand, allowing the sidewinder to travel over a kilometer in a night—quite far for such a small snake.

But this description of the way snakes move is of course crude. Attempts to describe their movements in terms of biomechanics, kinematics, and diagrams showing pivot points, arrows, and forces fall well short of reality. Snakes are agile, athletic creatures that move through their environment as it comes and on their own terms. To really see how a snake moves, just watch a gartersnake slide its way through your lawn next chance you get. To really know how a snake mounts a tree, you've got to stumble across a gray ratsnake going after a bird's nest, winnowing its way up the narrow grooves of a lichen-encrusted Florida live oak. To really know how a rattlesnake moves, you've got to find a five-footer crossing the road and fight the urge to run it over. Instead, watch how it works its way along so slowly that it hardly even seems to be there—using its belly scales like 200 individual little legs—before vanishing into the woods.

And if you want to know how a sidewinder moves, you've got to brave the hellish wastelands of the Mojave Desert—go somewhere with a somber name like the Devil's Playground, Funeral Mountains, Death Valley, Shadow Valley, or the Last Chance Range—and feel the raw, mummifying power of the sun, sand, faulted black rock, and salt.

Then you'll know.

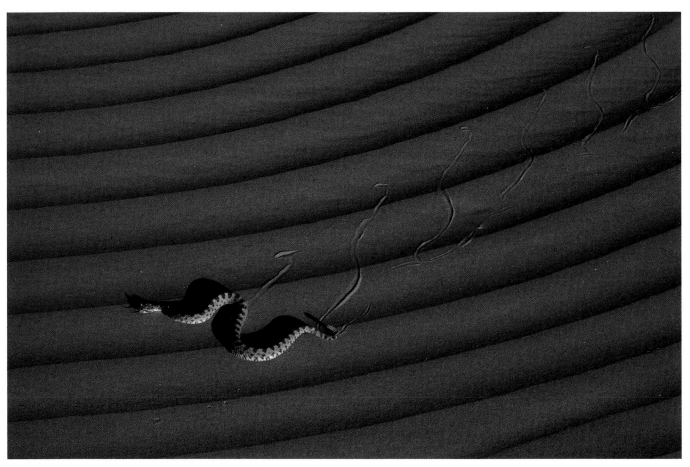

Sidewinder making tracks in sand. *Photograph by Bob McKeever*

3 · A Day in the Life of a Snake

Long ago—*hilahi'yu*—when the Sun became angry at the people on earth and sent a sickness to destroy them, the Little Men changed a man into a monster snake, which they called Uktena, "the Keen-eyed," and sent him to kill her . . . they hide now in deep pools in the river and about lonely passes in the high mountains, the places which the Cherokee call "Where the Uktena stays."

Cherokee tradition, JAMES MOONEY, *Myths of the Cherokee*

North Georgia Gold

The terminus of the Appalachians in Georgia presents their most dramatic view, because here they originate abruptly from the floor of a flat valley. The Great Valley of the Appalachians parallels the Blue Ridge Mountains from Georgia north all the way to Canada. Its most famous section is the Shenandoah of Virginia, but here the valley is called the Coosa and the mountains the Cohuttas. From Chatsworth, Georgia, you look east across a flat valley covered with productive pastures, and behold the dark green hulk of Fort Mountain, rising 2,000 feet into the sky. I was up there walking along a gravel trail, turning occasional logs and rocks. The forest had that oak-hickory forest smell: earthy, sweet, and spicy with nutmeg. It opened up ahead, and there was the wall. Large, flat slabs of gray, lichen-encrusted rocks were piled about at chest height in long rows. Most of the trees had only just begun leafing out, and the early spring sunshine warmed the rocks. The pattern of the wall didn't immediately make any sense—it wasn't square, triangular, or circular—and it vanished down the slope of the mountain to the west, north, and east. This was the rock "fort" of Fort Mountain—a mysterious archeological site that has perplexed historians and anthropologists for decades. Although most often thought erroneously to be some sort of fort made by Native Americans, there is no question that the views from the rocks would be commanding over a wide expanse of the Great Valley if the forest around them were cleared.

I began turning rock slabs along the slope near the wall. After a few minutes, I found what I was looking for. I lifted a large, warm, flat rock about half the size of a manhole cover, and coiled into a tight spiral underneath was a snake. It was a tiny and beautiful red-white-and-black banded snake—a baby eastern milksnake.

Its color was washed out by a gray film. The snake was about to shed, muting its colors. I gently rubbed the snake along its back, and the skin bunched up and peeled off. Then the colors were shiny and fresh, a living candy cane. I photographed the snake on its weathered rock home and released it.

A timber rattlesnake secluded on a southern Appalachian peak. *Photograph by Todd Pierson*

More special snakes would come. Later that afternoon some friends and I explored a remote outcrop 50 miles to the east. To the south, the sun was low in the horizon toward the rolling purple Piedmont foothills. This view is always hazy, sometimes even yellowish, mostly because of the poor air quality that seeps from the large urban bubble of Atlanta, a sprawling octopus about 75 miles to the south. The thick slabs of metamorphic rock ran along the side of the steep slope, forming narrow overhangs. Carpets of dark green *Selaginella*—a primitive, mat-forming clubmoss—grew in shallow pits along the bare gray-and-white marbled metamorphic rock. Clear water streaked black along the surface of the outcrop and gurgled from sphagnum moss seepages along its cracks. We made our way along the base of the diagonal rock slabs, our feet crunching under thick leaf litter, and scanned carefully for snakes. The leaves made a rhythmic *woosh woosh woosh*.

At the base of a small rock overhang we found a black timber rattlesnake in a tight coil, half-concealed in dappled sun. While we busily photographed it from a few feet away, the snake didn't budge. It didn't so much as flick out its tongue to acknowledge our presence. It just laid there in the leaves, soaking up the sun's warmth, and looked out with gray, unexpressive cat's eyes.

I drove the winding state roads back across the mountains after parting ways with my friends, in a terrific mood and barely believing the day's luck. I had seen two Appalachian snake specialties for the first time in the same day. I was speeding, making my way back down across the Great Valley. This region has always had terrific farmland—the rivers leaving the Blue Ridge carry rich minerals from the mountains and dump them out on the flats—and has therefore been the setting of sad human dramas and migrations. When Hernando de Soto marched an army through the region in 1540, the ancestors of the Upper Creeks lived in the area and built great ceremonial centers. De Soto's diseases and brutality caused population upheaval and declines, and the survivors moved downstream into Alabama. The Cherokees came out of the mountains and filled the void. There they adopted all the cultural leanings of the nineteenth-century South and became completely "civilized" in a shrewd attempt to assimilate and avoid persecution by Georgians. They had their own printing press that printed a native-language newspaper. They farmed land, built nice antebellum homes, and owned slaves. Then the first American gold rush—the Dahlonega gold rush of 1829—began in the Cherokee's mountain holdings to the east, and the influx of thousands of Georgians led to calls for Cherokee removal. They were forcibly removed in 1839 and sent west to Oklahoma on the infamous Trail of Tears. A final catastrophe came through the region in 1863–1864, when great armies in blue and gray played a deadly game of cat-and-mouse among these ridges and valleys, in what author Shelby Foote called a "red clay minuet."

The sun was going down as I hit the valley floor. I noticed a thin strip of bright tan along the middle of the road. Instantly I knew it was a snake—the way headlights reflect off a live snake is unmistakable—and I also knew that my lucky day had just got a lot luckier. I slammed down the emergency break, put my truck in neutral, checked behind me, and ran out and grabbed the snake. I was back in the truck, the snake in my left hand while shifting and steering with my right. I could barely control my excitement, because I'd just encountered a mole kingsnake for the first time in my life.

Mole kingsnakes are completely fossorial, or ground dwelling, and pursue fossorial prey, such as moles and shrews, by invading their underground burrows. They are a rich tan color with red playing-card diamonds on their back and a chessboard pattern on their belly. Their eyes are the same rich red color as the diamonds. They are relatively common in the Great Valley, a region with rich, loose soils, and, tellingly, a region of relict prairies. Owing to its long history of human agriculture, the valley has always been an avenue of open prairie-like habitat, and thus mole kingsnakes still emerge here on warm spring nights.

I found all three wonderful Appalachian species in a single lucky day. What each snake was doing when I found them might at first seem to be quite different. But in fact each snake was in a sense basking: the eastern milksnake under a warm rock, the timber rattlesnake in the open on a sunny outcrop, and the mole kingsnake on the warm pavement of a road just after dark. All three demonstrated an important aspect of the daily routine of snakes: maintaining their preferred body temperature. Each snake was also demonstrating one of the typical daily activity modes of snakes. The milksnake was doing what many snakes do for most of the time: hiding and remaining inactive. The timber rattlesnake was basking during the daytime. And the mole kingsnake was active nocturnally. These three gorgeous snakes of the Georgia mountains exemplified a representative selection of activities in the day in the life of a snake.

Consider a house cat. Or perhaps the family dog. What do they do all day? If they're like most cats and dogs, they curl up in a ball somewhere. As for cats, they often do it in some closet cubbyhole, and dogs might be found on the couch. They sleep all day. If some dramatic calamity happens inside the house, the cat's eyes flash open instantly, her claws like switchblades darting out, securing her position. If you get up to leave the house, the dog is usually up instantly, wagging his tail, to see if he can go with you. He is alert until he realizes he must stay at home, and then he goes back to sleep.

Now, consider where your pets would do these things if they were wild dogs or wild cats. In the case of the dog, he'd most likely dig himself an impressive burrow to curl into, to avoid oppressively cold or hot weather and to avoid being ambushed by other dogs or some bigger critter while sleeping. The cat would similarly occupy another animal's burrow, or, better yet, she would find some cavity high up in a tree to pass the day away in. Best to avoid critters that can't climb, namely, dogs. Both cats and dogs have similar activity patterns, and most predatory animals spend almost every hour of their day not spent hunting sleeping. Cats and dogs, like most predators, are lazy.

Snakes are similar. First, like cats and dogs, all snakes are hunters. The only exception to this characteristic is that a few snakes will eat dead things as well. But there are no snake vegetarians. We'll talk more about what snakes eat later, but for now, know that—like your family pet—snakes are born killers. Snakes are also lazy. Most vertebrate predators spend brief periods of time pursuing game or carefully planning and executing ambushes, and they spend the balance of their time resting. Carnivorous vertebrates do this to save hard-won energy; they can eat large meals and live off the carcass and the stored fat for days or even weeks. Snakes take this laziness to extremes that even the laziest dog could never surpass. In fact, there may be few other vertebrates on the planet that pass as much time being inactive as snakes do.

The similarities between snakes and your furry family pets soon end. As mammals, the dog and cats have a big metabolic furnace that they must keep constantly fired. The pilot light never goes out, and it must be constantly fueled either by occasional large meals or by many small meals. When cats and dogs rest, they can reduce the amount of fuel they burn, but they always have to squander a lot just to keep up their body temperature. As we have learned in chapter 2 about physiology, snakes don't have this problem. As ectotherms, their metabolic engine is much more efficient and practically runs on fumes. The snake engine is mostly solar powered. Snakes can achieve a body temperature nearly as constant and warm as mammals, and they don't have to burn up a lot of calories to do it. Instead, they use themselves like a solar panel, shifting into the sun when they get cool, and getting under cover when they get hot. Even better, they don't even need direct sunlight to get warm, but can find hidden little heat pockets under rocks. They can maintain their body temperature and still avoid enemies. For snakes, the environment is a veritable heat buffet to move through, to sample and select from, and to master.

This allows snakes to achieve supreme laziness: some snakes can eat one large meal *a year* and that's enough for them. The rest of the time they hole up somewhere, out of sight. So, the entire extent of almost any given day in the life of most snakes is spent in a burrow, under a log, concealed in leaf litter, under your porch, deep down inside a burned-out stump, under an old refrigerator, in the side of an arroyo, or up in a hollow in a tree. Given this, it is easy for many to assume that snakes are rare and occur at low population densities. The old adage is "snakes are where you find them"—a classic, dry understatement. There may be hundreds of snakes out there hiding and out of sight, and you'd never know it.

Snakes Are Secretive

With the advent of radiotelemetry studies on snakes, biologists gained the ability to find an individual snake no matter where it was hiding. Instead of those clunky transmitters you see mammalogists hang around the necks of polar bears, herpetologists use sleek packages that are surgically implanted

A juvenile timber rattlesnake hiding in a rock chamber. *Photograph by Bob Ferguson*

into the snake's body cavity. A biologist can relocate these transmitters with a radio antenna receiver. Telemetry studies show that, at least half the time, depending on their hunting style, snakes are under cover, hiding. If you walked through the forest, you'd never see them. The rest of the time they might be out, active or waiting in ambush, but even then, snakes are so camouflaged they're hard to find. This reveals a key and consternating feature of snake behavior: they're shy, and among the most secretive vertebrates. Their long, cylindrical body plan allows them to get into the tiniest hiding places, and, as anyone who's ever kept one as a pet knows, this characteristic also makes them exceptional escape artists. Even in well-studied species that occur at high densities—high enough to see many individual snakes in a day—we're only seeing a fraction of the snakes that are actually there. Some snakes have only been studied in special places where they are common; a classic example is the pygmy rattlesnake study at Lake Woodruff National Wildlife Refuge in Florida. Almost everything we know about this interesting little snake has emerged from this research.

Snakes have a tendency to go long time periods without being located. Scientists at the Savannah River Ecology Laboratory—which has been the site of constant and intensive study for 60 years—have only turned up two pine woods littersnakes during that whole period. This classic example demonstrates the secrecy of snakes. Snakes are where you find them.

Another reliable capture technique for snakes is the drift fence, a large sheet of metal flashing flush with the ground with buckets sunk at the ends. Snakes and many other animals, including mammals, like to slither or walk along obstructions and then go around them rather than try to go over or under. Snakes end up crawling along the fence and falling into the bucket, which is checked each morning. Checking drift fences always has surprises. You could walk the desert for decades and see only a fraction of the snakes you'd see if you erected a drift fence. I studied swamps in Alabama for four years, walking them intently one night a week and the next day from April to October, and I never saw an eastern hog-nosed snake, with the exception of a single individual I captured in a drift fence. Drift fence studies show that eastern deciduous forests are teaming with small littersnakes—small species like red-bellied snakes, earthsnakes, wormsnakes, and Dekay's brownsnakes. Occasionally you get lucky and find several on the surface, but they are always there whether you see them or not. Although snakes undoubtedly undergo population fluctuations, and are sometimes actually rare, a lot of their unpredictability is a result of their shyness, and it helps to use appropriate techniques to survey for them.

Active versus Ambush Hunting and Something in Between

Snakes eventually come out, and when they do, what are they up to? Practically all snakes undergo frequent, short-distance movements that are anything but straight. This was best demonstrated in the Mojave Desert, where the movements of coachwhips and sidewinders were determined not only by

Most of what we know about the pygmy rattlesnake comes from a series of studies conducted at the one place where they're common.
Photograph by Curtis Callaway

Bryan Hains

Farrell and May

Terry Farrell was out at Lake Woodruff National Wildlife Refuge with his three-year-old daughter when he found something that would change the course of his career. The refuge is a series of canals and levees dividing blocks of marshy wetlands and low hardwood hammocks thick with palmetto and anonymous evergreen shrubs. In winter, it provides a habitat for thousands of waterfowl and even has whooping cranes. Terry had been at Stetson University for over a year and visited the refuge dozens of times, admiring its abundant wildlife and, occasionally, its snakes. But that summer afternoon in the gathering twilight, Allison was with him, so he walked at a toddler's pace. That's how Terry first noticed a tiny salt-and-pepper-speckled snake coiled in the grass: his first pygmy rattlesnake. A couple of days later, he invited his colleague Peter out to see and photograph the snakes. They found several more, in what he described as "a mine field with little rattlesnakes everywhere." There is only one place where pygmy rattlesnakes are common enough to study, and Terry Farrell and Peter May—both ecologists with little snake experience—had found it. And it's a good thing they were the ones who did.

Farrell's previous research was on tide pools of the Pacific Northwest, where ecologists use the little pools as microcosms representing larger processes. They are easily manipulated and experimented with; you can remove all the predatory starfish and see what happens. Peter May had studied plant succession and birds, and butterfly foraging ecology. Farrell admits "it was a real jump for both of us," but attributes their success to "coming from a background where you need to do experiments." He also said that "established herpetologists were incredibly kind to us," because they, unlike ecologists, are motivated by "a love of the organisms," and were genuinely interested in what they would find.

So began an innovative collaboration examining the biology of pygmy rattlesnakes, leading to an unprecedented understanding of a little-known species. Terry and Peter were a great match intellectually. While Terry would "move through the environments rapidly—like a weasel," Peter "was more likely to see something cool and watch it. He had patience." Terry was "more of the idea guy" and would "play with the data," whereas Peter took meticulous notes and is a "really great writer." As a result, we know a great deal about pygmy rattlesnake population dynamics, diet, reproduction, and behavior. Terry and Peter coauthored dozens of papers and are usually collectively referred to by colleagues as "Farrell and May." They and their students conducted clever experiments, including my all-time favorite involving the scent trails of frogs. Eric Roth, then a master's student, made a frog scent extract, and spread little scent trails out in the jungly hardwood hammocks. As predicted, but still quite to their astonishment, the snakes began showing up along the invisible trails in their characteristic ambush posture, proving that pitvipers choose ambush sites based on the scent of their prey.

After two serious accidents with the rattlesnakes, Farrell can no longer handle them, but he continues to study the snakes with collaborators. Peter May perhaps grew bored of the snakes, and no longer participates in the studies. Instead, he enjoys the pleasures of natural history, engaging full time in wildlife photography. Terry seemed conflicted about this but was able to laugh it off. "I do miss that aspect of it, but it was kind of like a marriage. We were spending nearly every day with each other."

And I would be remiss if I didn't mention that Peter May did all those years of fieldwork on pygmy rattlesnakes from a wheelchair. In some ways this shouldn't matter, but as an inspiration to others, it really does. Terry remembers that "he'd be moving through thick habitat. It would be flooded and he'd be going around. It'd be muddy and he'd be going around. At some level I'd forget he even had mobility issues." Peter did not want to talk to me about the research and instead suggested I talk to Terry. I asked Terry whether he might be reluctant because he didn't want the wheelchair to be a concern. Terry thought that was possible, concluding, "That's the beauty of Peter."

Whit Gibbons
The Communicator

"Ain't it funny how much we don't know," Whit tells me through his fantastically toothy, permanent grin from the other side of a second-story porch in the 100-degree heat of his cabin in South Carolina. He interrupts our conversation a few times, calling back to his grandson Parker, who is up to his belly button in blackwater down in the swamp, switching between a dip net, fishing pole, kayak, and rope swing. Parker shimmied some 30 feet up a tree perched above 2 feet of water to adjust the rope swing, and Whit didn't so much as flinch. "Shouldn't kids be able to do these sorts of things?" he said with that smile again. He has a Southern accent as thick as the mud out back and sounds just like Strother Martin (from *Cool Hand Luke*: "What we got here . . . is a failure to communicate"). But his voice is smoother, and he's thoughtful, funny, and immediately strikes you as wise. More than any other living herpetologist, he is our communicator.

Like all herpetologists, he began his life as a kid too, a white-haired boy from Alabama who caught the herp bug after catching his first snake. He was assisting a Tulane graduate student by the time he was in junior high and went on expeditions with some of the legendary pioneers of our field at a young age. His career is a bridge that spans early twentieth-century biologists, his own cohort, and a half century's worth of students he has mentored.

Probably his greatest contribution to our understanding of snakes was the establishment of the Savannah River Ecology Laboratory in South Carolina. The location is a US Department of Energy nuclear reactor used to refine weapons-grade plutonium, but somehow Whit turned it into the longest-running herpetological study site in the world. For some snakes, the repeated captures of multiple individuals over the years has allowed a picture of their biology that would otherwise be impossible.

Whit especially likes techniques, some of which are explained in his terrific book *Their Blood Runs Cold*, which is a must-read for beginning herpetologists. He explains that for every reticent species there is some trick for catching them; you just need to invent it. So when we walked up the dry oak slope near his house, I cracked a smile when he proudly showed me his newest contraption. Three trees were clumsily wrapped with burlap held snug with bungee cords. It looked like some hillbillies' attempt to winterize a tree. The idea is that some lizards and snakes hide under loose bark, and by peeling back the bark you can find them. At first I dismissed the idea as nutty, quaint, and textbook Whit. But then I thought that maybe this is how all the tried-and-true capture techniques get started. Who would have ever thought that sinking buckets at the ends of a 1-foot-tall fence would turn out to be the most reliable way to catch snakes? He said his trees hadn't worked yet, but they only had to succeed once.

Parker walked up from the swamp smeared from head to toe with mud and asked for the sardines so he could bait the turtle traps. Grandpa tossed a can down. Then he tossed down his Swiss Army knife. Parker pried open the can and distributed the sardines among the traps, and while Whit was telling me which turtles they hadn't yet caught on the property, Parker tossed his pocketknife back up to Grandpa, who wasn't looking. I reached to catch it, but Whit turned and grabbed it before I could. He's still got it.

telemetry, but also by their tracks left in sand. Typical movements determined by telemetry are determined from point A to point B, and the distance is measured by a single, straight line. But the tracks of both species indicated they actually move in wriggling, tortuous paths that were 1.4 to 1.8 times longer than movements determined by ordinary telemetry. Long-distance movements also occur, but these movements are rare. Most snakes therefore move in frequent, short, convoluted pathways.

Much of a snake's activity pattern can be explained by whether it prefers to hunt actively or by ambush. Most American snakes can be conveniently divided into ambush hunters and active foragers based on their heritage. Most pitvipers are ambush hunters that lie in wait for prey at locations where they are likely to encounter them. Other American snakes are active foragers that move through the landscape in search of prey. There are a great deal more nocturnal active foragers than diurnal ones, and almost all diurnal active foragers are colubrids. Snakes with variable patterns—usually snakes that are diurnal during the cool weather of spring and fall, and become nocturnal during the heat of the summer—are about as likely to be ambush hunters as active foragers.

The daily movement patterns of snakes can be difficult to interpret. Snakes using either ambush or active hunting styles can have similar movement patterns; they both undergo many short movements and only occasionally move long distances. Active hunters move longer distances than ambush hunters, but why would both kinds of snakes show such lopsided movement patterns, with shorter movements made more frequently than longer ones? Attempting to comprehend such movements, trackers of northern pinesnakes in Tennes-

A collection of small terrestrial snakes found during the spring. These snakes are common in eastern forests but are not always easy to find. *Photograph by Bob Ferguson*

Coachwhips are active hunters that move rapidly through their habitat, often with their heads held high, visually searching for prey. *Photograph by Ian Deery*

Sidewinders are ambush hunters that lie in wait—often half-buried in sand—waiting for prey. *Photograph by Rulon Clark*

see expressed uncharacteristic bewilderment in a scientific article, noting that "it is unclear why snakes travel long distances from an area containing a plethora of underground retreats to get to habitats resembling the area previously occupied."

Hunting snakes are probably similar to mammalian predators like weasels. Such carnivorous mammals forage widely within their home ranges along familiar old hunting paths: checking burrows, ripping open logs, pressing their snouts down holes, and giving brief chase when they discover a small mammal or other potential meal. This pattern is referred to as "trap lining," named after the habit of human hunters setting traps along a route and returning to check them every day. Similarly, snakes—both ambush hunters and active foragers—move among patches of potential prey, checking out the area to see if they get lucky and find a meal. For true active foragers, this is a nearly continuous process, and movements are frequent and seldom interrupted. They search for prey and frequently give chase when they find it. But few snakes behave like this (examples of continuous active foragers are all diurnal and include coachwhips, racers, whipsnakes, speckled racers, and perhaps also patch-nosed snakes). Ambush hunters do this to adjust from one ambush point to another, perhaps because they didn't get any "bites," or because they did not smell high concentrations of potential prey at a previous location. Almost all of our pitvipers forage this way most of the time. This leaves a large number of American snakes that do not fit well into either category.

Most American snakes are what can perhaps more appropriately be termed "raiders," as they frequently move short distances while probing small burrows, nests, and other retreats looking for prey. They do not ambush prey in the sense that pitvipers do—they do not lie in wait hoping for prey to simply walk past. Instead, they actively search for prey. But they do not engage in pursuit in the sense that true active foragers do, because they are usually not fast enough to catch their prey. The reason why a Tennessee pinesnake moves from one series of burrows to an equally fine series of burrows a kilometer away is that it has checked the first burrows for food, and, finding no rodents, it moves on to the next one. Having checked the fresh burrows, it might then return to the first burrows to try its luck again. A prairie rattlesnake moves short distances and takes up ambush sites under a sand sage, occupying the entrances of rodent burrow, facing out. After a few days waiting, a rodent never appears, so it moves to another copse of sage a dozen meters away. A prairie kingsnake moves underground among small mammal burrow complexes, tunneling from one burrow complex to the next. It emerges only occasionally to crawl atop the surface grasses to check another series of burrows somewhere else. A black ratsnake in Maryland moves along the ground and throughout the canopy of trees searching for rodents in burrows and bird's nests along branches. After successfully raiding a blue jay's nest, it moves 100 meters out of the forest to hide under the foundation of a nearby house to digest its meal. A week later, it returns to the same tree and surprises a nesting wood thrush. An eastern kingsnake in the New Jersey pine barrens crawls about the forest floor only 21% of the time, probing the roots at the base of trees as it looks for snakes, rodents, or eggs. Between searches it stays coiled among the shallow underground roots of a pine.

The time of day when snakes hunt is largely dependent on what the snake is hunting. Ambush hunters must lie in wait for moving prey. Active foragers and raiders search for immobile prey. Long-nosed snakes crawl across desert roads at night on their way to dig sleeping lizards out of their burrows. Long-nosed snakes are nocturnal raiders. Northern pinesnakes cruise across large areas, descending into burrows looking for sleeping pocket gophers and other rodents, cleaning out the nests when prey are found. Pinesnakes are diurnal raiders. Sidewinders lie half-buried in sand under a creosote bush at night, with their head and neck cocked and ready to strike a passing kangaroo rat. Sidewinders are nocturnal ambush hunters. Eastern diamondback rattlesnakes lie

An eastern ratsnake on the prowl at night. *Photograph by Zack West*

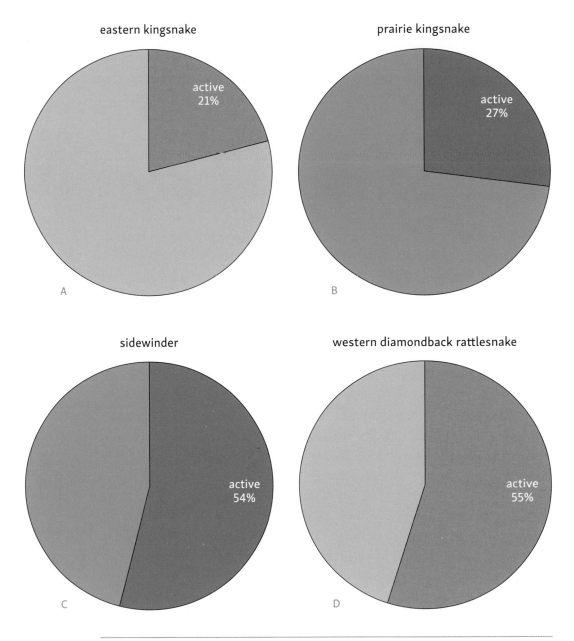

eastern kingsnake

active 21%

A

prairie kingsnake

active 27%

B

sidewinder

active 54%

C

western diamondback rattlesnake

active 55%

D

Activity patterns of representative hunting strategies and daily activity patterns. Blue charts represent nocturnal species; green charts indicate diurnal ones. *A and B*, active hunters that forage widely looking for prey; *C and D*, ambush hunters that lie in wait for prey.

along the edge of grassy rabbit tunnels coiled in the same fashion. Eastern diamondbacks are diurnal ambush hunters. Finally, Sonoran whipsnakes cruise the scrub, jerkily scanning for lizards. Whipsnakes are diurnal active foragers. Some snakes use a cunning combination of tactics, depending on the time of day or year: cottonmouths are usually nocturnal ambush hunters, but sometimes they hunt during the day, and sometimes they actively forage along the banks of swamps day or night, hoping to surprise a stationary frog or snake.

I don't mean to imply that every time you see a snake out in the open it is hunting. Sometimes a snake crawling across your backyard may be moving for any number of inexplicable reasons. Perhaps it's undergoing a seasonal migration, or perhaps it was the victim of its own hunting strategy and evicted from its burrow by a raiding weasel, ringtail, or raccoon. It takes a little practice to distinguish whether an ambush hunter is in an ambush pose or simply coiled and inactive. Usually ambush hunters are fairly loosely coiled, with their head and neck in an S-shape, and usually they're facing some sort of perpendicular object that serves as a runway for their prey. Cottonmouths set up ambush sites along the water's edge. But sometimes you'll find cottonmouths in a rather tight coil right next to the water's edge, or even facing *away* from the water's edge. Sometimes you find cottonmouths in loose coils far from water facing the base of shrubs. What are these snakes up to? Sometimes we just don't know.

Thermoregulatory Behavior

Many of the inexplicable movements and sudden appearances of snakes can be explained by thermoregulatory behavior. Many snakes spend a great deal of time basking when they are on the surface, actively lying in the sun to elevate their body temperatures. Digestion and gestation both occur more quickly and more efficiently at higher body temperatures. Basking is also frequently associated with shedding, and basking is most often observed during springtime. Snakes often choose characteristic sites for basking. In the deserts and open interior of the western United States, snakes are often found basking near rock outcrops. In the swamps and fens of the eastern United States, cottonmouths and massasaugas choose naturally elevated platforms, like a grass clump or shrub hummock. Watersnakes are prone to basking upon narrow branches along the water's edge, so that they can quickly drop out of danger into the water. Snakes don't even need to come out on the surface to thermoregulate. Gartersnakes will select refuge sites under rocks with certain dimensions to maintain their preferred body temperature. Rocks too thin and small are not selected, because it gets too hot under those. You won't find a gartersnake under a rock that is too thick and big, because it's too cool. Rocks of a medium size and thickness are just right, allowing the snake to maintain its preferred body temperature in safety and under cover.

Snakes use two thermoregulatory strategies. Some snakes simply allow their body temperature to conform to the environmental temperature. For example, when they are on the surface, pygmy rattlesnakes in Florida lie in wait to ambush frogs most of the day and most of the year, even in winter. Their average body temperature matches the average local temperature almost exactly, and if you measured their body temperature during the day, it would go up and down with the air temperature. Other species are obsessive about maintaining a steady temperature, moving erratically through their environ-

A brown watersnake basks in a snag over a stream. *Photograph by Pierson Hill*

Timber rattlesnakes barely expose themselves while basking under warm rocks.
Photograph by the author

ment to maintain it at a set level. Coachwhips might start their day stretched out to face the desert sun, spend the morning ducking through mesquite chasing lizards, and coil in a cavity in a dusty arroyo during the midday. Then they're out again in the dappled shade of shrubs, and they might even cruise an open creosote flat in the late afternoon, going back to the same arroyo to spend the night. During a day in which temperatures may fluctuate as much as 26°F, the coachwhip's body temperature may hover at temperatures higher than human body temperatures, and may vary by just a few degrees. At night, it might cool down to close to the soil temperature, requiring a few extra minutes the next morning to heat back up.

Sidewinders and coachwhips in the Mojave Desert have thermal characteristics associated with their hunting style, and these are broadly applicable to many other kinds of snakes. Sidewinders spend a good deal of time active on the surface in their characteristic ambush "craters," allowing their body temperature to fluctuate widely during the day. They move infrequently and are cryptically patterned, and have the typical stout body proportions of ambush hunting snakes around the world. They rely on camouflage because

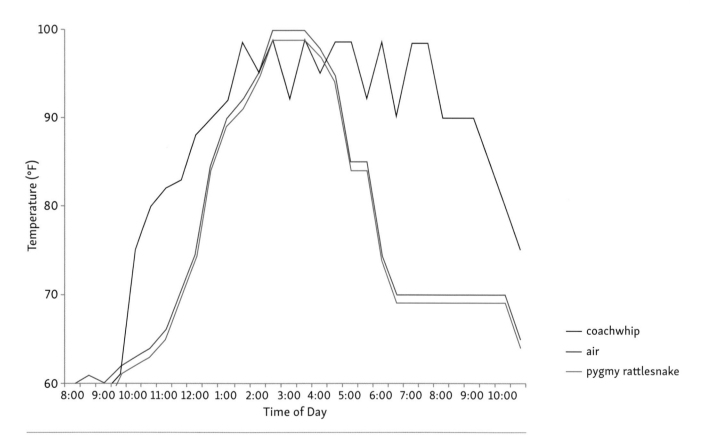

The thermal profile of an actively thermoregulating coachwhip versus a pygmy rattlesnake content to conform to the environmental temperature.

they remain motionless on the surface for long periods of time, making them vulnerable to predators.

The same desert flats are home to actively foraging coachwhips, which undergo frequent, brief movements associated with hunting. They spend more time under cover than the sidewinders. Their body temperature remains within narrow confines during their daily routine. When they begin to overheat, they quickly dive into a rodent burrow to cool off. Then they move back out again. They are long and skinny, not particularly camouflaged, and rely instead on their speed to avoid enemies. The sidewinders and coachwhips of the sun-drowned desert flats of Southern California represent the extreme ends of the continuum of snake lifestyles. Most American snakes operate somewhere between these extremes.

Factors That Influence Daily Activity Patterns

Snakes' daily activity patterns vary owing to all kinds of factors, such as geography, habitat, season, moon phase, or weather. Some snake enthusiasts become devoted interpreters of these factors in order to find snakes, and like soothsayers they consider each in turn. They might wet their finger and raise it into the air, detect the local wind speed and humidity, consider the incline and aspect of yonder ridge, or with great confidence choose the perfect rock to turn, which then presents absolutely no snake. We all have our stories of great finds, and we all have our suggestions for when to go looking for snakes. None of us can seem to remember all the fruitless hours looking with no success when conditions were just right. For example, studies have tried to establish a correlation between moon phase and snake activity, with some suggesting that snakes stay under cover when the moon is full, and others suggesting the opposite.

Although some factors certainly influence snake activity patterns, they are by no means reliable enough to be effective predictors of their behavior. And for many uncommon snakes, there are simply too few observations to determine what kinds of environmental factors influence their activities. Geography and the related influence of elevation can influence snake activity patterns, mostly owing to great differences in temperature across latitude and altitude. Gartersnakes in the cool, temperate rainforests of the Pacific Northwest are strictly diurnal and must bask frequently as they come in and out of the cool water they inhabit. In the Southeast, gartersnakes can be diurnal but are frequently found on roads at night as well. On the other hand, activity patterns can be surprisingly consistent from one place to the next. Racers in Utah, Kansas, California, and South Carolina all spend a similar amount of time being surface active, with some slight differences. Racers in completely different habitats—from the Colorado Plateau to the tallgrass prairie, from coastal chaparral to thick second-growth eastern forests—all share similar habits. These racers are even different colors, the western ones being a greenish yellow, and the South Carolina snakes being black. Despite this, the racers do similar things, and their body temperatures are usually kept at a remarkably uniform average of 89°F to 91°F.

Activity patterns and habitat use can vary seasonally. In the Sonoran Desert near Tucson, three kinds of rattlesnakes show different patterns. Throughout the year, including winter, these snakes are above the surface on average about half the time. Surface activity and movement are highest in all three species during the wet summer monsoons, when blue thunderheads cruise across the Arizona sky, and Tucson becomes as humid as Memphis. During the monsoons, they spend up to 20 hours of each day on the surface, most of that time alert in their ambush positions. During most of the year the three species utilize the entire overlapping landscape of washes, creosote flats, and rocky outcrops, but during the monsoons the three rattlers hunt in separate habitats. Tiger rattlers perch in clefts along the arroyo, now occasionally rushing with water and filled with the mechanical calling of canyon tree frogs. Western dia-

A tiger rattlesnake in Arizona's Sonoran Desert. *Photograph by Timothy A. Cota*

mondbacks fan out into the barren, monotonous creosote flats, which fill the air with the penetrating smell of wet telephone-pole grease. Black-tailed rattlesnakes occupy the rugged bajadas among the barrel cactus and giant saguaro.

In summary, many factors can influence a snake's decision to come out to crawl about the surface, including its feeding, reproductive, or shedding status; its hunting or thermoregulatory strategy; evolutionary heritage; climate; geography; habitat; and season. Because these factors are so variable and most snakes spend a majority of their time hiding, it's a wonder we know anything at all about snakes.

Armed with this explanation for the daily routine of snakes, we can better attempt to understand what the mountain snakes mentioned in the beginning of this chapter were doing. The eastern milksnake was probably basking, albeit under a rock. It would be difficult to determine whether the young snake was truly basking or completely inactive. Yet the fact that the rocks were warm suggests the former. Many snakes can achieve their preferred temperatures without ever having to venture out into the open. Also, spring basking in snakes is often prolonged over a period of weeks (as we shall see in chapter 4) and accompanied by shedding, as was the case with the milksnake. The timber rattlesnake was likely basking (this snake was also about to shed), and probably would have spent the rest of the day next to the rock before retreating under the outcrop at nightfall. The mole kingsnake was certainly active, but exactly what it was up to is more difficult to interpret. It may have been getting warm on the pavement in preparation for a night of raiding burrows, or it may have been migrating from a rocky outcrop to its summer foraging grounds.

Who knows? To better appreciate the lives of snakes, we next examine their seasonal activity patterns.

4 · A Year in the Life of a Snake

I saw a striped snake run into the water, and he lay on the bottom, apparently without inconvenience, as long as I stayed there, or more than a quarter of an hour; perhaps because he had not yet fairly come out of the torpid state. It appeared to me that for a like reason men remain in their present low and primitive condition; but if they should feel the influence of the spring of springs arousing them, they would of necessity rise to a higher and more ethereal life.

HENRY DAVID THOREAU, *Walden*

A Cottonmouth Year

The seasons of a Georgia swamp can seem virtually nonexistent. If visited at any point from April to October, the untrained eye may discern no changes. The days are long and hot, and the entire active season seems to be one long summer—the genuine, buoyant spring and crisp autumn inexplicably missing. Compared to places farther north, seasonal changes take place at a leisurely pace, but upon closer inspection, and after achieving a more intimate familiarity, one realizes that anywhere you go on the planet—owing to its slight tilt toward the sun—there are always constant and sprinting upheavals, eruptions, and cycles. To get a feel for seasonal changes in Georgia, you have to practically get down on your hands and knees and crawl face-first through the rich muck, looking down into its swirling clay waters at least three or four times a week. Then you understand that the place does have seasons, and contained within that languid summer there are in fact the hurried activities of a thousand creatures contributing to the year's changes. Cynics might instead categorize the seasons of Georgia solely in terms of its biting insects: winter is sand fly season, spring is mosquito season, changing to deer fly season, gnat season, and so forth. But instead I focus on the more abundant species that pack their activities into the long southern summertime, resulting in—for those with the patience to observe it—a net gain in seasonal activity.

I got to know these seasons well when I disappeared into a Georgia swamp for nearly a decade studying its snakes. I anticipated the emergence of snakes with as much esteem as singing frogs, blooming wildflowers, and arriving birds. I occasionally returned to civilization for dinner and a movie, for classes and graduations, for birthday parties and work. But those years in my memory are now an amalgam of yellow-green summers catching cottonmouths.

For the cottonmouth, the new year begins in March, when the first days of spring are bright enough to sunburn you. Big purple clouds propped on narrow shafts of sunlight in February give way to white cumulus clouds aloft on warm air. Giant W- and V-shaped flocks of trumpeting sandhill cranes slowly drift north, their 5-foot wingspans little more than specks flying high among jumbo jets. Cottonmouths appear at the base of rock outcrops along the creeks, next to overhanging banks of roots, atop the hummocks topped by small trees and grasses, and circle old stump holes in the hardwood forests, never too far from water. On warm days they can be found this way, curled in a tight coil, gray and pitiful looking.

For a few weeks, dozens of cottonmouths can be found near their wintering sites basking before descending into the floodplains to begin hunting. In April, before a riot of yellow-green vegetation fills the landscape, equal numbers of cottonmouths can be found basking as those lying next to narrow waterways, with their heads and necks cocked and aimed, ready to ambush prey. They are hungry after their winter slumber.

Then the trees leaf out, their hundreds of big trunks acting like a thousand straws sucking down the winter water table. The water sinks by at least 2 feet, and archipelagos, peninsulas, and a network of channels appear from the yellow water, forming a perfect maze of hunting edges for cottonmouths. For the

A cottonmouth lies in wait for prey in a Georgia swamp. *Photograph by Noah Fields*

next three months, the swamp is the setting for biological dramas rivaling those of the African savannah. Cottonmouths are a constant hazard for anything in the swamp smaller than a raccoon. Frogs, fish, birds, rodents, snakes, crawdads, and numerous other creatures live in perpetual anxiety that the next root they encounter may in fact be a cleverly disguised snake. Along the quagmire of cutoff channels, oxbows, sloughs, and beaver projects, dozens of cottonmouths are propped along the edge of the water, patiently waiting.

The hot days pass slowly, and you'll be tempted to think that time has arrested, that the year has stalled in a long, humid purgatory of summer. Red-eyed vireos sing a drawling, repetitive warble from the treetops. Acadian flycatchers call abruptly from the hot, green gloom under the canopy, as if cursing the heat. Cottonmouths soon pile atop driftwood and hummocks, basking half in the filtered dark green, half out in the dappled summer sun. Such snakes have a lump in their belly, and are basking to quickly digest their meal before returning to the hunt.

Late summer—the breeding and birthing season—is the most mysterious and exciting time for observing cottonmouths. Half the females each year will give birth, and half are available for mating and will give birth the next year. The snakes seem to disappear. Many are likely under cover and up to sexual deeds. From July to September, you are equally likely to find a hunting cottonmouth as you are to suddenly stumble upon a quartet of plump, pregnant females basking atop an upturned umbrella of windthrown roots. Or you may turn the corner of a swamp channel and suddenly discover a female snake half out of her old skin—as if undressing—attended by a male lustfully nudging her back with his snout, passionately darting his tongue along her sweet, musky back.

Late summer progresses to a brief autumn, when the oppressive summer humidity at last yields to a mercifully crisp, dry heat. Nights cool off, and one by one the annual vegetation finishes, producing fruit, some edible to man, some only edible to rodents and sparrows. Cottonmouths become scarce. Females give birth to a small litter of colorful, yellow-tipped babies on hummocks during the first week of September and stay with them—perhaps to defend them—until their first shed a week later. As the first frost threatens, the snakes make their way back to winter dens, often to the same stump hole or granite outcrop used the previous year. In November, the sandhill cranes pass again—their fluting cries whispering down from the heights—and the snakes have assumed the same temperature as their winter quarters, reducing their heart rate to perhaps one beat every few hours, just to keep the blood going.

Winters are a short and rather exciting time in a Georgia swamp, and only a few days pass when no plant or animal activity occurs. All the knotweeds and sunflowers that stood head-high in the marshes during the summer now lay pushed down like a tangled mat, making a perfect home and providing food for swamp sparrows and orange-crowned warblers. The woods provide gleanings for small flocks of titmice, chickadees, nuthatches, and kinglets. Solo hermit thrushes and mobs of waxwings search for every last late-summer fruit. The sap never stops flowing down here, so sapsuckers hammer away at perfectly round sap wells all winter long. The cottonmouths stay in their holes and sleep.

It's difficult to develop a detailed impression of the annual cycle of snakes. You need a decent-sized population so that you can reliably encounter individuals almost every time you go out to look. For snakes this is difficult. Either the numbers are too low and dispersed, so that discerning annual patterns takes years of study, or the snakes are only available for a short time period, so a wider seasonal picture is elusive. For these reasons, we know little about the seasonal activity patterns of such snakes as long-nosed snakes, which are fairly common but widely dispersed across the western deserts. We even have an incomplete picture of the well-studied common gartersnake, since they are abundant near their hibernation dens during spring emergence, but then disperse widely into the surrounding habitats for the rest of the year. By comparing numbers of snakes encountered seasonally across decades, we can piece together general seasonal patterns for many species. Radiotelemetry studies have given us excellent insights into the seasonal activities of larger snakes. But for most species, especially some of the interesting small ones, we have only vague notions of their seasonal patterns.

Fortunately, accumulated information about several well-studied species has given us a good picture of the annual cycle of many American snakes. The major events of this cycle are spring emergence, migration, summer hunting, mating (covered in more detail in chap. 5), and hibernation.

Emergence

Throughout the American countryside each spring, snakes unfurl from their winter quarters on warm days and begin their annual cycles. This happens on schedules determined by lengthening of the days, thawing of ice and snow, and warming of the ground: the same schedules that determine the migration of birds, choruses of frogs, and emergence of insects. The primary trigger is ground temperature, which must warm up enough to reach the underground retreats of snakes. The exact timing of emergence therefore varies from year to year depending on how warm a spring it is. The wave of activity travels up the American countryside from the south a few days at a time, and up the sides of mountains at the same rate. Peak spring emergence occurs around St. Patrick's Day in the Deep South. Snakes begin emerging as early as late March and early April in the New Jersey pine barrens, mid-April in eastern Colorado, late April in northwest Ohio and southeast Michigan, and mid-May in Wyoming and Canada.

Where frosts are rare or absent, snakes can be active year-round. Their activity levels are reduced during the cooler months, but they never enter true

A gartersnake emerges from its winter quarters with snow still on the ground.
Photograph by Bob Ferguson

hibernation. Ratsnakes as far north as Waco, Texas, can be found crawling around in January. In that same month, an eastern diamondback rattlesnake on the Georgia coast might be found stretched out in the low but sufficient sun with a small rabbit in its belly. Black pinesnakes in Mississippi, canebrake rattlesnakes in South Carolina, and eastern kingsnakes in south Georgia all have lower activity levels in winter but do not completely cease movements. Western diamondback rattlesnakes can be found basking outside rock shelters on sunny days between rains of the Sonoran Desert winter. Perhaps the most active winter snake is our largest: the eastern indigo snake continues moderate activity and even breeds during the winter months.

The general boundary for complete winter seasonal inactivity is above the 32nd parallel of latitude (above the southern half of South Carolina, Georgia, Alabama, Mississippi, Louisiana, Texas, New Mexico, and Arizona). And there are only two places in the United States where snakes continue activity essentially uninterrupted. South Florida and Texas approach the Tropic of Cancer below the 28th parallel. Snakes there remain active except during the few brief periods of cool weather that occasionally affect those latitudes. In fact, the period of greatest activity in south Florida is during winter, when wetlands become dry and snakes there are prone to prowling around.

Many snakes are easiest to find during emergence, and in the northern states, snakes can be quite common during the spring. Perhaps the densest population of snakes on the continent is across the US border in Canada, where tens of thousands of gartersnakes emerge from rocky fissures during the first few weeks of May. Similar but less astounding emergences at snake "dens" occur throughout the Northeast, the northern prairies, and in the mountains. Multiple snake species will often den together during the winter, and they emerge together in sizeable numbers.

Many snakes remain in the neighborhood of their hibernaculum, or over-wintering site, for some time after emergence. Male gartersnakes and water-snakes emerge before females and wait for females. When they emerge, their brief spring mating period commences. This emergence pattern is associated with mating; for prairie rattlesnakes, which do not mate in spring, males and females emerge simultaneously and immediately leave the vicinity of the dens. In many snakes, spring emergence is associated with a prolonged period of basking. Because the body temperature of snakes rises quickly in the spring sun, this one- or two-week period of basking does not seem to be directly associated with thermoregulation per se. Instead, their body may need a period of sunbathing to prepare physiologically for the active season. Snakes that reproduce in the spring probably use this basking period to rapidly initiate or conclude the growth of sperm or eggs, which is a temperature-dependent process.

Great Basin rattlesnakes emerging from a rock outcrop in spring. *Photograph by Bob Hansen*

A smooth greensnake basks during the spring. *Photograph by Ashley Tubbs*

Copperheads found under warm tin in the spring. *Photograph by Noah Fields*

The spring basking period is a convenient time to search for snakes, even those that hibernate solitarily in southern locations. At this time they can often be found near rock outcrops and stump holes, and under boards, bark, rocks, old refrigerators, lawnmowers, barn doors, and other debris. And spring is when the snake hunter's "tin sites" are most productive. Some snake enthusiasts go through the trouble of gathering up tin slabs from blown-over barns and distribute them in old fields in hopes of cultivating snake basking sites. It is not necessary for a snake to lie in the open to bask. Especially during emergence, many snakes get warm while remaining under cover. Corrugated tin sheets are ideal basking sites, and therefore ideal places to look for emerging snakes.

Migration

After basking, mating, or immediately after emerging from their den, snakes migrate to their hunting grounds. For many snakes, this is a straight-line movement to a habitat completely different from their winter den. Gartersnakes in Canada crawl a couple of miles in just a few days, making their way steadily and directly to marshes where they spend the summer. Racers, whipsnakes, and prairie rattlesnakes in Utah slither in a purposeful, direct pattern for about a week as they spread out from their rocky dens. They then switch to a more irregular movement pattern: circling around, zigzagging, and shifting from one area rich with prey to another. Massasaugas in Colorado migrate 2 miles across the prairie to sandhills, where their prey is common. Long movements

Massasaugas in Colorado migrate from rock outcrops used for overwintering to sandhills used for hunting. *Photograph by Curtis Callaway*

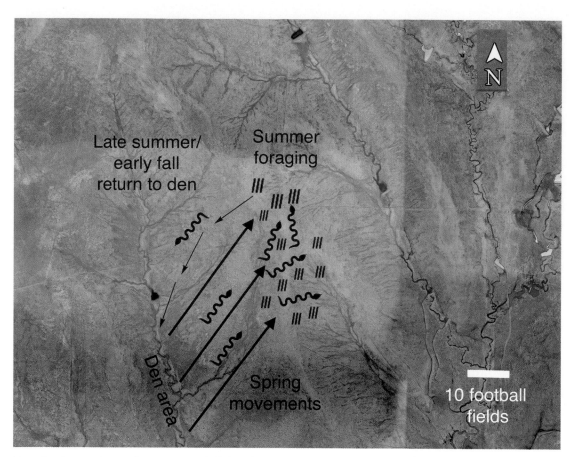

Late summer/
early fall
return to den

Summer
foraging

Den area

Spring
movements

10 football
fields

The migratory pathway of massasaugas in Colorado.

are not required for some species; cottonmouths and watersnakes in southern swamps often hibernate in uplands and make short movements downslope to the wetlands where they spend the summer. But many simply take up winter residence in stumps, vegetation clumps, windthrows, muskrat mounds, and old beaver lodges, and they are already in place when spring arrives. Similarly, desert rattlesnakes occupy small mammal burrows, rock outcrops, and arroyos, and they do not have far to go when the scrublands become warm. Mojave rattlesnakes occupy any convenient mammal burrow when it gets cool, and simply stay there until the weather warms up again a few months later.

Home Range and Movements

The seasonal activities of snakes are best determined by radiotelemetry studies. These involve a single researcher following anywhere from six to over a dozen snakes throughout their activity period and locating them perhaps once or twice a week. The implanted transmitter allows the researcher to find the snake no matter where it goes—up in a tree, concealed within a brush pile, a few feet underground in a rodent burrow, across a stream. This also means that to find each snake, the researcher has to be able to go where the snake leads them, no matter what obstacle may appear and no matter how rough the country. The tracker's work is thus long and lonely, and quite difficult at times. Their capacity for translating the minute clicking beeps of the receiver into a distance and a direction seem to me more like necromancy or water divining. I've only been out tracking snakes a few times, and even for the tourist it can be drudgery. Once in coastal California tracking southern Pacific rattlesnakes, we spent a cool misty morning finding snakes, one of which evaded our tracker because it was coiled under his boot the entire time he tried to pinpoint its location. To locate another snake, we had to wade through a 5-acre stand of chest-high poison oak. For the next week, we suffered from tender, itching blisters. I've always admired the skill of the snake trackers, and especially the information that can be generated by these studies. The snake trackers know more about their snakes than anyone.

Early on, no clear generalities emerged from these studies. At first it was unclear whether snakes have true home ranges—areas they use and have an intimate familiarity with—or whether they simply move about the landscape randomly. Some snakes do move randomly—watersnakes in Wisconsin emerge from rock dens along lakesides and simply move haphazardly along the shoreline among cattails feeding on fish and frogs. The only focal point of their movement is the hibernation den, which they return to each fall. The areas they use each summer continue to expand every time they are located by the tracker, rather than reaching some set boundary. Similarly, brown watersnakes cruise along the Savannah River in South Carolina feeding on catfish, occasionally and quite randomly moving upstream, downstream, and across the river. They are probably only motivated by their success in capturing catfish and by their hunger when they fail.

But dozens of studies conducted since the advent of surgically implanted transmitters show conclusively that most snakes move within a well-defined home range. After emergence, snakes use the landscape in predictable ways, often returning over and over again to the same hiding places, indicating they have familiarity with their surroundings. Most snakes can be relocated hundreds of times and their home range does not continue to expand, indicating their ranges have boundaries. Snakes tracked from year to year reoccupy a home range with the same general size and shape, perhaps shifting a little here and there but remaining for the most part steady over time. These home ranges are often represented as polygons showing only the outline of the locations farthest along the periphery of the snake's wanderings.

The sizes of snake home ranges are influenced by many factors and can vary widely between species and even within the same population of snakes. Bigger snakes have bigger home ranges; the smallest known home range of an American snake is that of the eastern wormsnake, whose average body size is just around 25 cm, and whose average home range (0.025 hectares; a hectare is a metric unit that is equal in area to about two American football fields, or 2.5 acres) was determined from a Kentucky population. The eastern wormsnake is also the smallest species that has ever been tracked. No radio transmitters are small enough to implant into a snake the size and width of a no. 2 pencil, so researchers tracked them by attaching radioactive tags and followed them with Geiger counters. Considerable advances in radiotelemetry will be required before we're able to track additional small snakes. The largest home range determined for an American snake was from a population of indigo snakes in south Georgia, which range over an astounding average of 359 hectares.

Even though they have well-defined home ranges, most snakes do not actually have territories. Territories are home ranges that are defended, and there is little evidence that snakes defend the boundaries of their home ranges from other snakes. This has so far been documented in only one type of snake from Taiwan. Kukrisnakes will briefly declare sovereignty over the space around a sea turtle nest and defend it from other kukrisnakes. This way, they can feed leisurely on the entire nest of over 100 eggs, one at a time, over a period of several weeks. This they do using their namesake fangs, which resemble kukri blades in shape. The eggs are larger than the snake's head, and they become lodged in the back of the mouth. There the kukri teeth slit them open, and they collapse and their contents are swallowed. The teeth are also used to jab rivals who approach their defended nest.

Most snakes have no such means or motivation to defend territories, and the majority of snakes instead show the hallmark of an undefended home range: their home ranges overlap extensively with their neighbors. Only a few

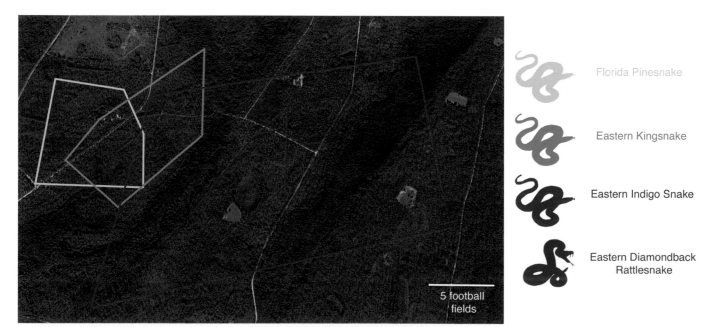

Florida Pinesnake

Eastern Kingsnake

Eastern Indigo Snake

Eastern Diamondback Rattlesnake

5 football fields

Average home ranges of four snake species in a longleaf pine forest of southern Georgia.

Natalie Hyslop
Tracking Endangered Species

I saw my first indigo snake out tracking with Natalie Hyslop, who did her dissertation research on indigos in south Georgia. Natalie, a self-identifying tomboy, grew up wild, exploring the creeks and woods near her house in Marietta, Georgia. She was encouraged by her parents to take part in such iconoclastic behaviors, igniting a "lifelong fascination with nature" that would ultimately place her among the most respected snake trackers in the country. She tracked federally threatened copperbelly watersnakes for her master's research, and was now tracking the legendary indigo at Ft. Stewart, a huge military base garrisoning the 3rd Infantry Division.

We walked out along a sandhill on a hot October afternoon, and Natalie explained the nuances of interpreting a telemetry signal. The signal can ricochet off subtle imperfections in

Natalie Hyslop demonstrating radiotelemetry.

terrain and make the process maddening at times. I asked her what it takes to be a good tracker, and she told me that plenty of practice helps, and that "you have to be stubborn." But she pointed out that some people simply never get it. And she was sure to mention that it is *hard*. She tracked a copperbelly out into an open wetland, trying to finish for the day as an April blue Kentucky thunderstorm dropped on her. Wood ducks suddenly exploded from the water—just one memorable "heart attack" moment she had out tracking. She's been inadvertently shot at twice, discovered the camp of an AWOL soldier, and stumbled within 50 feet of what must have been an entire battalion of the US infantry in full armament. But she knows these kinds of obstacles come with field research. She told me plainly, "You quit. Or, you bring your game. It speaks to the passion you have for the field." And as far as the art of telemetry goes, she figured it out on her second day.

We walked down a slight ravine to a dark bay swamp parting the sandhill. She began pointing out some of the trees common in bay swamps. For Natalie, the snakes are fascinating, but the bigger ecological picture is even more exciting. She told me with evident passion that the most memorable part of her research is that it is "such a privilege to see an entire ecosystem go through its seasonal changes."

She pointed out the snake, and I was able to briefly watch its large body slither back into the jungly swamp. It moved under a pile of driftwood and vanished. But Natalie would find it again, over and over again, no matter where it went and no matter how far it crawled. For years. Her study ultimately revealed some extraordinary insights about indigo snakes: how they move back and forth between upland and swamp habitat, and how they can move incredibly large distances. Hers are the largest reported home ranges for any snake.

snakes show nonoverlapping, or exclusive, home ranges (eastern kingsnakes are a notable exception), and so far it is impossible to say whether in these cases the snakes actively defend the boundaries of their home ranges. But snakes that have exclusive home ranges may be an indication that some snakes might at least avoid the boundaries of other snake's home ranges. Coachwhips in Southern California do this when they are packed together in small habitats. In a large, fairly pristine habitat fragment, coachwhips have large, nonoverlapping home ranges, while in a tiny fragment they have small home ranges that overlap extensively. This indicates that, at least in areas with large amounts of habitat, coachwhips would prefer to keep their space, while in small habitat fragments they are forced to share the landscape.

Home range size is also frequently associated with the availability of prey. Colorado massasaugas move long distances from winter hibernacula before settling into small home ranges where their prey are abundant. Prairie rattlesnakes in Wyoming follow a similar pattern, moving among patches of sagebrush habitat where rodent populations are locally abundant. Black pinesnakes and northern pinesnakes have different home range sizes, but both snakes move about the landscape searching for patches of their preferred prey.

Within their home ranges, snakes have definite preferences for certain habitats. They don't simply move randomly within their home ranges and use all the habitats within it proportionally. Instead, they spend most of their time within certain habitats, often in far greater proportion to their availability,

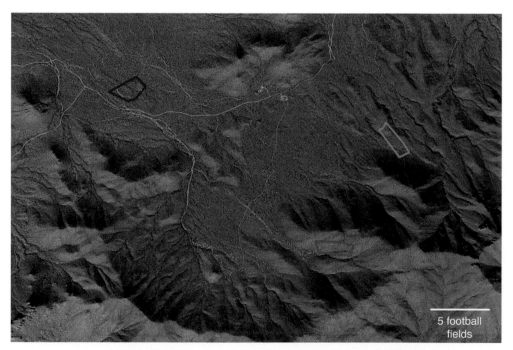

Western Diamondback Rattlesnake

Tiger Rattlesnake

Western Black-tailed Rattlesnake

Average home ranges and habitat tendencies of three rattlesnakes in the Sonoran Desert of Arizona. During the summer rains, western diamondbacks tend to occupy creosote flats, while tiger rattlesnakes occupy arroyos and black-tailed rattlesnakes occupy bajadas.

and avoid certain common habitats that do not provide food, cover, or basking sites. This means that, within their home range, most snakes only use a small fraction of the total area. Wisconsin bullsnakes spend a majority of their time among the bluffs arising from creeks, which only make up about 4% of the total land area. While certain natural habitats are favored, snakes, including bullsnakes, usually show a distinct avoidance of agricultural fields. That is not to say snakes are never associated with human-modified habitats, because snakes frequently use landscapes like old pastures and even use human dwellings and debris as retreats. But snakes usually avoid wide-open monocultures of soybeans, alfalfa, and other crops.

Habitat preferences determined by telemetry don't always match the typical descriptions from naturalist's impressions and books, and have revealed unexpected surprises about the lives of snakes. Perhaps the most famous example of this was Howard Reinert's study of Pennsylvania timber rattlesnakes. Until this study, most scientists thought that the preferred habitat of timber rattlesnakes was open rocky outcrops. But telemetered timber rattlesnakes quickly radiate out from the outcrops into thick, brushy forests nearby. This is where they spend the majority of their time, lying in wait next to logs, waiting for their prey to approach. Nobody knew timber rattlesnakes did this, because nobody could find timber rattlesnakes camouflaged on the forest floor without the help of telemetry.

Mating

Snakes have distinct mating seasons and behaviors, which are discussed more thoroughly in chapter 5. For now, I only mention that mating is another primary influence on seasonal home range size and seasonal activity in snakes. Generally, seasonal peaks of activity and movement are associated with mating, and many of these apparent peaks of activity are associated with increased observations of males. Males often have much larger home ranges than females, and this is most obvious during the mating period. At this time, male snakes undergo determined and often straight-line movements in search of females; straight-line movements may be best to intercept females when they are dispersed widely and quite unpredictably in the landscape. This habit has been noted in prairie rattlesnakes during their late-summer mating period. Female northern watersnakes move more frequently during their period of maximum reproductive readiness, making it easier for males to track them.

In snakes that mate in the spring and late summer, the home range typically expands twice. Many snakes mate in the immediate vicinity of the hibernaculum, but some snakes mate in spring after they disperse from the hibernaculum, and large male home ranges are most prevalent at that time. Females sometimes undergo long straight-line movements in the summer, but this has been observed only in species that lay eggs; they often undertake long migrations to nesting areas. Eastern hog-nosed snakes move rapidly and in a straight line in June toward sandy dunes along the shore of Lake Ontario to locate the loose, warm sand needed for nesting. Western foxsnakes make similar movements during July to lay eggs in chambers on raised levees in Illinois marshes.

Hibernation

After the activity period of hunting and mating, snakes return to hibernacula for overwintering, which occurs as early as October in northern states and as late as November or December farther south. Their movements back to the hibernacula are similar to spring migrations; they move directly and in straight lines, suggesting they have intimate familiarity with their surroundings. They do not retrace their trail from their summer and spring range, and instead move directly overland back to the hibernaculum, indicating they have onboard navigational abilities. Newborn snakes have the capacity to follow scent trails of other snakes to find an appropriate hibernation site. A newborn timber rattlesnake follows the precise track of its mother to her hibernaculum. Baby massasaugas retrace their tracks back to the open rookery sites where they were born, which provide them with access to good burrows for hibernating.

Snakes hibernate for the same reason that mammals do: food is scarce during the American winter. It is better to wait out the hard times by lowering their metabolism and activity until food is plentiful again. Many of the prey

A mountain patch-nosed snake in autumn will soon return to its winter den. *Photograph by Zack West*

eaten by snakes are exceedingly scarce during the winter, so snakes hibernate partially for this reason. In addition, for snakes, body functions are nearly impossible at low temperatures, and the risk of freezing is much greater for snakes than for birds and mammals. For this reason, most snakes hibernate well below the frost line, deep underground where frost and freezing temperatures do not reach.

The farther north you travel, the more likely snakes can be found hibernating in large numbers in specific winter retreats. Thousands of red-sided gartersnakes hibernate in limestone fissures in Manitoba. Fewer, but still impressive, numbers (200–300 individuals) occupied a cistern in Wisconsin. By contrast, gartersnakes in southern locations hibernate in shallow burrows alone. Several species often coexist in the same hibernaculum, and this is also more common farther north: red-bellied snakes, western wormsnakes, Dekay's brownsnakes, and smooth earthsnakes occupied the same ant tunnels in Kansas; the same cistern in Wisconsin contained hundreds of gartersnakes as well as foxsnakes and red-bellied snakes; rock outcrops in Utah housed hundreds of prairie rattlesnakes, gophersnakes, racers, and striped whipsnakes; sinkholes in the Colorado prairie contained desert massasaugas, prairie rattlesnakes, racers, bullsnakes, coachwhips, western hog-nosed snakes, milksnakes, and plains gartersnakes; a south Alabama sandstone outcrop sheltered small numbers of eastern diamondback rattlesnakes, timber rattlesnakes, racers, copperheads, and coachwhips.

Only certain dens allow snakes to survive the harsh winters, and even in long-established hibernacula, winter mortality can be high. This is usually due to prolonged periods of freezing temperatures, or from warm-blooded predators that unearth and eat the snakes during hibernation. In the southern United States, snakes have more options to pass the shorter winter in comfort, so they are more widely dispersed and typically hibernate alone and in shallow dens. There are exceptions: massasaugas in southern Ontario do not den communally and simply make do with tunnels made by small mammals and other holes in the forest. Likewise, communal dens are known for some southern species; small numbers of cottonmouths will emerge together from old root holes, and rocky outcrops serve as dens for timber rattlesnakes and eastern diamondback rattlesnakes in the sandy coastal plain of Alabama.

How snakes select their wintering sites is somewhat mysterious; attempts to divine patterns from the slope direction, vegetation cover, solar orientation, and habitat structure of preferred dens are rarely successful. Similar rock outcrops with the same features can often be found nearby and have no snakes emerging from them at all. Conversely, northern Pacific rattlesnakes select den sites within rocky outcrops on south-facing, gradual slopes. Even the rock sizes they use are a distinctive, intermediate size; not too small, which would encourage cave-ins, and not too large, which might allow entry of too much cold air. These perfectly sized and placed rocky outcrops were chosen significantly more often than random sites close by. Many dens in far northern locations share such features, especially the southern slope aspect. Southern slopes are warmer than others because they receive more sunlight, and these may be especially sought after by snakes if they undergo a period of spring basking.

Overwintering sites for snakes in southern latitudes can be simple rodent burrows not much deeper than a foot underground; such shallow nooks will remain above freezing all winter and keep the snakes hidden. Perhaps the most coveted hibernation sites of the South are stump holes; the roots of diseased or fire-killed trees rot away and leave deep cavities easily accessed by snakes. In the great longleaf pine ecosystem along the southern coastal plain, frequent fires burn out the undergrowth and maintain a grassy, park-like aspect. Occasionally the fires kill hardwood trees and even the fire-adapted longleaf pine. Fires eventually burn out underground taproots of snags—following them deep underground and smoldering for weeks—carving out deep stump holes used by many species.

A great variety of dens have been used by American snakes: the tunnels of ants, crayfish, rodents, gopher tortoises, desert tortoises, and prairie dog towns; old wells, cisterns, and homestead foundations; human trash piles, mulch piles, and brush piles; broken-down barns, old brick bridge pilings, and old cars; rocky fissures in karst landscapes with sinkholes; limestone, sandstone, and granite outcrops; under logs, rocks, leaf litter, sand, and other surface debris. The commonality is that snakes will use just about any underground retreat for hibernation, so long as they don't have to dig it themselves. Contrary to folk belief, snakes don't typically make their own holes. Only a few kinds of American snakes frequently dig their own burrows, and all three have special features that enable them to do so. The northern pinesnake has a cone-shaped head and flattened snout scale used for this purpose, and it arches its neck into a "J" shape to scoop out dirt. Eastern and southern hog-nosed snakes have sharp, upturned snout scales used for digging for their prey. All three species are known to dig themselves a short way underground for hibernation.

Many snakes prefer hibernacula that allow access to groundwater. This may be so that they can maintain a body temperature that is slightly warmer than freezing (since flowing groundwater rarely freezes) and also remain hydrated while hibernating. Even terrestrial snakes like gartersnakes, red-bellied snakes, and foxsnakes have been found hibernating completely submerged in an old cistern. They don't drown because their metabolic, breathing, and heart rates are all minimal. Likewise, timber rattlesnakes seek out seepages emerging from forested ravines and wedge themselves into these creases of flowing water to hibernate. Many snakes therefore emerge from hibernation covered in mud, and their skin does not take on its revitalized appearance until after their active season's first shed.

Specific hibernacula are often used year after year by the same individuals. Snakes using Utah rock outcrops showed a close affinity to their hibernacula from year to year. They returned to the same rock outcrops between 90% and 100% of the time, despite the fact that some of the outcrops were just 1 km away from others. Gartersnakes have been seen crawling past perfectly suitable spots used by other snakes to get back to their own preferred dens. A remarkable long-term study of northern pinesnakes in New Jersey gives us unparalleled insight into the hibernation habits of this species. Specific dens were monitored for 26 years to determine how often the snakes came back to the same dens, and how often certain dens were used or abandoned. The snakes used the same sites for as many as 16 years! Astonishingly, some snakes would return to the same dens after a 10-year hiatus, during which time the den went unused by any snake. In some cases this was because dens were dug up by skunks, foxes, or people. But in other cases the reasons why they abandoned the dens were a total mystery.

Within the den, snakes do not necessarily stay still. The temperatures within most hibernacula can be as high as 40°F to 50°F even though the temperature above ground is below freezing. Snakes are able to move around at such temperatures and can move farther back into the depths of their den if the temperatures get too cold near the surface. Some snakes actually prefer to cool down to low temperatures during the winter hiatus, because this causes their metabolic rates to plunge to imperceptible levels. It is even possible that snakes lose too much weight if they are too warm during the winter; cottonmouths in Virginia lost more weight and suffered more winter mortality in warm years compared to cold years. Many snakes require cooling periods for proper physiological function; the only way to induce sexual behavior in red-sided gartersnakes is to cool them down for a few months and then warm them back up. Snake breeders often subject their snakes to artificial hibernation in captivity for the best results. When kept at appropriate, cool temperatures, snakes lose little weight during winter dormancy, usually only around 10% of their body weight. The little energy they lose from moving around during the winter is supported by metabolizing stores of body fat.

Within the dens, snakes remain capable of slight movements and retain their ability to escape temperatures below freezing. Unlike mammals—which enter a pitifully vulnerable state during hibernation—many snakes are immediately capable of defending themselves and attempt to escape when surprised by predators or people in their dens. There is a general tendency for snakes to move to slightly warmer temperatures within their retreats as temperatures above ground plunge, and they move toward the surface again when surface temperatures begin to warm up. After a long period of dormancy, ranging from six months in the North to three months in the South, the snakes emerge from their winter hibernation and begin a new annual cycle.

Joanna Burger

Pine Barrens Naturalist

Joanna and her consummate colleague Bob Zappalorti checked the tag number, and checked it twice. They checked it a third time and were forced to admit they made a mistake. All her pinesnakes receive passive induced transmitter (PIT) tags—small devices the size of a Tic Tac candy. You can determine its unique number with a barcode reader, much like they have at the grocery store. They had just captured and released the snake but wrote the wrong number. They'd have to find it again. They walked back to the same spot, and the snake had vanished. After combing the area for 30 minutes in an expanding circular grid, they found no sign of the snake. Then Joanna looked up, and to her astonishment the pinesnake—usually considered a terrestrial species—was 30 feet up in a pine tree. Undaunted, Joanna—then perhaps in her fifties—climbed the tree, leaned out among the swaying boughs, and got another reading from the tag.

Most of what we know about pinesnakes comes from Joanna and Bob's long-term research in the New Jersey pine barrens. They began studying them decades ago and not only revealed extraordinary details about their habits, but also some unique insights about snakes in general. For example, their studies of nesting and hibernating in pinesnakes are practically all we know about these important activities in snakes. They investigated the nest chambers of pinesnakes—their dimensions, depth, characteristics, and fate. Many were dug up by skunks, and some were eaten by shrews. They even found a scarletsnake eating the contents of one egg. An alarming number of eggs were found completely missing, with human footprints leading from the scene of the crime. Their hibernation studies are the most extensive ever attempted—Joanna has relocated the same hibernating snake in 19 out of 23 years.

Joanna grew up on a farm in Upstate New York and quickly realized she "wasn't going to be happy picking tomatoes." Instead, she indulged her love of the natural world, and was banding birds by the time she was 10. She discovered her first gartersnake while exploring the irises and rock garden near her farmhouse. She said catching snakes "was going along really well until I brought back a copperhead. That didn't go over well."

Joanna is now a distinguished professor at Rutgers University. She's written dozens of books and scientific papers, and her research has prevented uncontrolled development of the pine barrens, including the installation of yet another Walmart. She finds it "most rewarding and challenging to find out how people and wildlife can survive to the benefit of both. It's a constant battle. But it's worth it." Her attitude is upbeat and boisterous, her accent that of a pleasant New Yorker, reminding me of my parents. In reference to her research, she says things like "It's so exciting," "It's so fascinating; it's a puzzle," and "it's just wonderful." I'm certain she'll continue studying her pinesnakes for years, and thanks to her, our understanding of snakes in general will expand. But when it comes to climbing trees, let's hope she begins delegating such tasks to students.

Prior to winter thunderstorms, during a walk through the uplands border-ing a Georgia swamp, you may find a cottonmouth just outside its burrow in January, enjoying a brief warm snap. It won't have to wait long. After a few more weeks, the tiny flowers of Elliot blueberry appear, the yellow puffballs of spicebush burst, followed by the rich, sweet smell of 400,000 red maple flowers atop the floodplain canopy. Add to this the hypnotic scent of yellow jessamine—the envy of any high-end perfume maker—and you know that a new cottonmouth year is set to begin anew.

5 · Snake Sex

Mating Balls

Rick Shine and his longtime collaborator Mats Olsson sped across the Manitoba prairie in a rental car, racing the sun in early May 1997. The prairie is glacier-flat, interrupted only by southeast-running marshes and thickets of skinny aspens, and the wide arc of the sky was already turning that North Country rose purple. They were just 10° of latitude south of the Arctic Circle and one hour by commuter flight from Churchill, where you can see polar bears during the winter. They were supposed to get in after dark, have dinner, and wait until morning to start. But they couldn't resist stopping in at the study site. They got out, tested the air, and doubled back to get their coats on. It was cooling off fast in the fading light. They raced down the trail among bare, enamel-white aspens and thick prairie grass. The first thing they noticed was a strange, pervasive rasping sound. In the dim twilight ahead, they saw a depression in the limestone.

Then they saw 10,000 snakes in a pit the size of Rick's living room.

Rick is Australian, and the most accomplished herpetologist of his generation. He studied lethal red-bellied blacksnakes in the Blue Mountains of New South Wales and water pythons in the monsoonal marshes of the Northern Territory. He has published nearly 900 research papers, most of them on snakes. But until that evening he had never seen so many snakes in his life.

The red-sided gartersnakes of Manitoba's Narcisse Snake Dens are perhaps the most thoroughly studied snake population in the world. They are so important I've made an exception here to talk about Canadian snakes. Upon emergence in spring, they mate frantically over a period of a couple of weeks. They are so eager to mate that they ignore human observers. When he heard about these dense concentrations of snakes, Rick had to get in on the action. Rick once said, "It's a good thing you Yanks have gartersnakes, or you wouldn't have anything to study."

Red-sided gartersnakes are nearly black with a solid yellow stripe down the back. They are the prairie version of the common gartersnake of the eastern United States and valley gartersnake of California; it is the northernmost representative of this common, widespread species. When active and breathing heavy, their inch-thick bodies expose red checkerboard patches. They are bright eyed and alert. Now imagine 10,000 of them, all seemingly identical; many writhing around in a solid mass; others "zipping around like reptilian dynamos looking for a date."

Rick said, "that first glance around showed me more snake courtship in 10 seconds than I had seen in the last 30 years." The strange rasping they heard was the sound of thousands of churning, tumbling, rough-scaled snakes.

A mating ball of red-sided gartersnakes.
Photograph by Tracy Langkilde

Rick and Mats stared slack-jawed at the masses of snakes, and they "just squealed, and occasionally grunted to each other." Rick has a big smile with an impish twinkle, and I can imagine him there, running around picking up fistfuls of snakes. I know that uncontrollable pit churning around in your guts when you find something thrilling, and it makes you yip. I wonder if nonscientists have ever experienced this emotion. Rick said, "my inner five-year-old was in ecstasy."

In contrast, I can count the number of times I've seen the cottonmouth mating—the snake I've studied the most—on one finger. This was not for lack of trying. The one observation of courtship I saw came before I even began research on these snakes, and I just stumbled upon it by luck. One afternoon I was walking along the floodplain and about to cross Morning Creek, a 10- to 15-foot-wide tributary of the Flint River. The river floodplain is a kaleidoscope of different-aged stands of bottomland timber—recently drowned beaver ponds with dozens of snags; older ponds thickening with red maples and edges grown with willows; ancient, dark stands of swamp chestnut oak, green ash, overcup oak, and swamp hickory. The creek meanders through this jungle patchwork of swamps and marshes, sandy bottomed and tea brown when running clear, silty and yellow when flooding. The old stands have numerous windthrows, the dead trees lying on their sides and roots pulled out, lying vertical. I came around a windthrow and found a pair of cottonmouths lying on the creek edge. The male was on top of the female, nudging her almost tenderly with its snout. The female on bottom was halfway through shedding. I only got to watch for a few seconds before the male noticed me (you can almost see a look of surprise in their eyes, despite the lack of eyelids) and quickly swam to the other side of the creek. I subsequently studied cottonmouths for seven years, visiting the swamps at least once per week throughout the active season, and never saw anything like it again.

Given this, you can imagine why the incredible mating congregations of red-sided gartersnakes in Manitoba were such a boon to researchers. But it does make me wonder just how applicable this knowledge is to other species, since such densities are not typical. Red-sided gartersnakes spread out into the surrounding prairie during the summertime and return to their crucial rock dens for the winter. Because they are at the northern boundary of the distribution of all North American snakes, only certain rock shelters will do. This leads to absurd population densities in the spring. Most snakes never achieve such dense population densities and are distributed rather widely across the landscape. And we know that the densities of animals can greatly influence their mating habits.

But the gartersnake studies and others have given us a surprisingly good understanding of the reproductive biology and behaviors of a number of American snakes. These insights have been determined using techniques ranging from elegant experiments to cutting-edge genetic technology. From

Rick Shine

The Biggest Dog in Herpetology

Mats Olsson

Rick probably doesn't remember this, but I was feeling on top of the world after defending my master's thesis, so I wrote Rick an e-mail declaring him to be the "biggest dog" in herpetology and announcing that I wanted to work with him for my PhD. I assumed he must receive hundreds of emails from prospective students, so I attempted to distinguish myself by being as informal as possible. He responded politely and rebounded all my jokes, but informed me that it would be tough because of the Australian scholarship system. I eventually worked with him but not in any official capacity: he paid for my visit to Australia to work with one of his students on Cane Toads. I ended up marrying her.

Rick really is the biggest dog in herpetology. He has published more papers about snakes than any other person, living or dead. He collaborates with so many people that he has written papers about snakes not only from his native Australia, but also New Caledonia, Taiwan, China, Southeast Asia, Europe, the United States, and Canada. At this point his reputation and productivity have spiraled into a whirlwind force of their own: people seek him out to help maximize the impact of their findings. Although most of his research involves Australian snakes, he has done enough work in America to deserve his own place in this book.

His success is a result of a few noteworthy advantages: number one, he is clearly brilliant. Not just smart, but also creative, inquisitive, efficient, and funny. He is a prolific writer; he can quickly turn out a manuscript or help sculpt someone else's work. His students often receive thorough revisions within hours of sending him a draft. When I worked with his group in Australia, they were artists at setting up studies and shrewd data hoarders. The results are clear: unparalleled insights into the recalcitrant biology of snakes.

this research it is becoming clear that some of the most interesting and outlandish facets of the lives of snakes happen in the bedroom.

It is perhaps fitting that an animal with such Freudian connotations should have bizarre and iconoclastic sexual habits. Puritans beware! What follows is an unflinching account of mating and reproduction in snakes: transvestite males who seductively attract their rivals to receive a free massage and then ripple their bodies upon receptive females; May Day orgies with a cast of thousands; females who fertilize their eggs at their leisure using 10-month-old sperm; nightmarish, four-headed reproductive organs covered with spines; females who prefer mating with multiple males; gelatinous chastity belts that glue shut the female orifice; deceptive females masquerading as males to select victorious bulls; 8-hour sex marathons.

Perhaps the one consolation I can provide the pious will only enrage puritanical readers further, owing to its audacity, unexpectedness, and impertinence: rattlesnakes are good mothers.

Finding Mates

During the mating season, male snakes crawl around looking for females using their vomeronasal sense. This is the chemical sense associated with the forked tongue and the special sensory organ in the roof of the mouth. Females release pheromones signaling their sexual attractivity—special chemicals build up in their skin and are released. This is not unlike the estrous behavior of dogs "in heat." Female northern watersnakes become sexually attractive soon after shedding, when pheromones in their skin reach high levels. Blood levels of estrogen also spike at precisely this time, just as in mammals during estrous.

A mating aggregation of plain-bellied watersnakes. *Photograph by John Hewlett*

Shedding is usually associated with estrous in snakes. Females may also engage in another behavior that makes them more available to males: female northern watersnakes move more frequently during estrous. This behavior, combined with the tendency for males to increase their movements during the mating season, helps snakes intercept one another for mating.

The mating season of American snakes varies greatly, but some general patterns are known: most colubrids breed in the spring, soon after emergence from hibernation. Snakes that mate during the spring can take advantage of the proximity of males and females near hibernacula at this time. When more widely dispersed, it may be difficult or inconvenient to rendezvous. Pitvipers often mate in the late summer and fall, and some species mate in both the spring and the fall. It is unclear what advantages these variable mating seasons give the snake. Fall mating may actually force males to search widely for females, with only the best males successfully finding them. In many species the precise mating season is unknown or ambiguous because there are so few observations of mating to determine it.

Reproductive Cycles

The ovarian cycle of snakes is slightly different than in most mammals. Instead of ovulating tiny eggs like mammals, reptile follicles undergo a prolonged period of yolk provisioning. The protein-rich, fatty yolk is produced by the mother from stores in her fat bodies and liver. The developing follicles grow quite large and are ovulated at a much greater size than in mammals. The production of estrogen and the sex pheromone correspond with yolking, so at some time during this process, estrous occurs as well.

All American snakes ovulate during late spring and early summer, regardless of the yolking or mating pattern. This means that in all species, eggs are laid around June and develop over the summer. For species that give birth to live young, they are ovulated and retained in the oviducts until birth, usually in middle to late summer.

Mating Systems

Mating systems are the different ways animals breed, in reference to the inevitable inequity that arises between males and females in terms of who breeds with whom. These systems range from monogamy—which is most common in birds, where both sexes are needed to help raise a clutch of eggs—to various kinds of polygyny, where males mate with multiple females. Polygyny is common in lizards and mammals, which are often territorial, with dominant males guarding a harem of females. Males tend to mate with more than one female if they can, since sperm is cheap. Only in cases where their help is needed to raise their own young do males stick around with a single female. Because they must invest heavily in eggs that are energetically expensive to produce, the best way for females to increase their reproductive success is to mate with the best male they can and to take good care of their young. So, in general, females are choosy about whom they mate with.

The mating balls of red-sided gartersnakes are a classic example of a mating system called scramble competition. In animals with short breeding seasons and high population densities, hundreds of individuals make up a mating aggregation, and males swarm over available females in an attempt to mate. During May in the Canadian prairies, males begin emerging from rocky fissures first. Females then emerge one at a time, resulting in perhaps the most skewed sex ratio of any vertebrate mating system: as many as 5,000 males may vie for a single female. Such extreme orgies can result in suffocation and death. Males with good endurance that are quick to find females, skilled at competing with other males, and fast at mating tend to have better mating success. Strangely, certain males also behave like females, and even release feminine pheromones. These "she-males" end up attracting their share of males. This appears to be a devious strategy for quickly receiving a high dose of body heat from their swarming rivals. The she-males at the bottom of the pile are also safer from marauding crows.

Females briefly lose their attractivity after they mate, and this is reinforced by a bizarre chastity belt secreted by males. In addition to sperm, male red-sided gartersnakes secrete a "sperm plug" that glues the female's cloaca shut for a time. That way, the male's sperm has more time to travel and fertilize more eggs in the event that the female mates again.

Gartersnakes south of Manitoba participate in less dramatic mating balls; a population in Wisconsin takes part in much more dignified mating balls of only thirty males to one female. Perhaps even more typical are the mating balls of watersnakes, which involve far fewer individuals—from a chaste one-on-one sex ratio to perhaps as many as six males courting a female. This mating system is also considered to be scramble competition; males emerge from hibernacula first, females emerge one at a time, males follow and find her, and dozens of males court her aggressively. In this system it is unclear whether the female has any way to choose a male.

In pitvipers and some colubrines, male dominance is established through ritualistic battles, and male reproductive success is increased by finding and mating with multiple females. This mating system has been dubbed "mate-searching polygyny," which emphasizes the male tendency to search widely for females during the mating season. This mating system is typical of animals that are not aggregated and instead widely dispersed throughout the environment—like most snakes.

This mating system is typical of prairie rattlesnakes in Wyoming, which hibernate in small mammal burrows (often in prairie dog towns) and emerge during the high-country spring, when patches of snow may still be present among the sagebrush. They then move to feeding areas, which have clumps of vegetation and high densities of rodents. They remain there until late summer,

Mating ball of Dekay's brownsnakes. *Photograph by Bob Ferguson*

Combat "dance" between male western diamondback rattlesnakes. *Photograph by Brian Sullivan*

at which point males and females are spread all over the prairie. Mating begins in late August and early September, when the morning air becomes crisp and elk squeal like giant violins in the mountains. Then male prairie rattlers double their home range and move in straight lines in an attempt to intercept females. If they encounter a male, they engage in spirited, writhing, full-body wrestling matches, but these are rare in Wyoming. If they encounter a female, they court her insistently, and, if she allows it, they mate. Males that can find more than one female will increase their mating success.

Early naturalists witnessed pitvipers engaging in mating "dances," where two snakes make contact, rise up, and vigorously entwine the upper halves of their bodies. Occasionally one snake slams down the other in what appears to be an arm-wrestling match using the whole body. At first this was assumed to be a form of courtship, but eventually it was determined that such dances were between males. A female snake was often found as a spectator close by. It then became clear these were some form of male-male combat for access to females. Larger males usually win the fights, which can go on for hours. The victor is able to topple his challenger more often because he can raise himself up slightly higher, gaining hooking leverage with his neck, allowing him to throw down his opponent. At some point the loser is toppled one last time, so he gives in and attempts to escape. The dominant male then chases him from the scene, doubles back, and begins courting the female in the vicinity. Losers typically lose subsequent fights, even if he fights with a smaller male. This is a result of the "loser effect" driven by high levels of stress hormones. This well-known response has even been documented in humans; people whose favorite sports team loses a big game have elevated stress hormones. Surely you've experienced this agony of defeat yourself.

For copperhead losers, it is much worse. When a loser is presented to a female, she arcs her neck vertically in the air to face him. This behavior is identical to the initial phase of the male combat bout. Because losers avoid subsequent fights, they do not accept her challenge, and instead crawl away. Winners presented to females are faced with this same gesture, and quickly raise themselves to accept the challenge. The females then demure, lowering their head back down submissively. The male then begins to tenderly court her. In this manner, females have developed a canny way to distinguish dominant males from losers; they mimic the behavior of dominant males so they can ward off losers and mate instead with winners. The stress hormones in losers remain high for almost a week, and during this time they are effectively removed from the breeding pool. Male-male combat has also been documented in colubrines such as cornsnakes, ratsnakes, and gophersnakes, although their mating systems have not yet been studied extensively.

American snakes have long been characterized as having scramble mating systems or polygynous mating systems. But a radical proposal threatens to upturn our view of the mating systems of snakes. This was largely possible

Combat between male red cornsnakes. *Photograph by Crystal Kelehear*

owing to the availability of state-of-the-art genetic paternity analyses of the clutches of various wild snakes. These show that, almost invariably, females lay clutches sired by more than one male. This observation, coupled with the large number of snakes whose breeding system includes multiple males courting a single, large female, has led some to suggest that perhaps our view of snake mating systems was unduly influenced by the gender of the researchers themselves. Until recently, most snake biologists were men, and it should be admitted that we are a macho bunch. When such men witnessed mating balls of gartersnakes, it was natural for us to assume it was a free-for-all competition among males to be the one snake to fertilize the female's eggs. When we witnessed rattlesnakes gripped in combat, it was natural for us to assume the mating system was polygynous, where dominant males who fought best got the most females.

While still controversial, there is likely some credence to the idea that mating systems of most snakes are not polygynous. Even in well-studied scrambling gartersnakes and watersnakes, females mate with more than one male. Even in pitvipers, where males engage in masculine combat to win females, females mate with multiple males over the breeding season. The two sexes are caught between competing motivations: males try to inseminate multiple females to increase his mating success, and for obscure reasons females also mate with multiple males. Snake mating systems are more appropriately defined as promiscuous.

Although it is obvious why males would want to mate with more than one female, what reason would females have to mate with more than one male?

A large female brown watersnake accompanied by courting males. *Photograph by Pierson Hill*

There are drawbacks to mating repeatedly; female lizards have developed interesting colors and behaviors to signal males they have already mated and avoid harassment by males. Mating is a vulnerable time when snakes are susceptible to predation. If a single male can fertilize her whole clutch, then a single mating should be enough.

In other animals, the advantages of females mating with more than one male (polyandry) are often vague or involve factors that cannot explain the mating system of snakes. It is often assumed that females mate with more than one male to increase "genetic diversity" or to introduce "good genes" to her clutch. In birds, polyandry is rare, but in the cases where it occurs the females leave their clutch for the male to attend, so it is directly advantageous for her to mate with many males to increase her reproductive output. They drop off a series of clutches to several dutifully caring males. Astonishingly, the females of these bird species are more brightly colored than males, fight other females for access to males, and actively court their male nannies—a complete sex reversal from the usual pattern. But in snakes, the males do not attend eggs, so the advantage for female snakes is quite mysterious.

The only vertebrate with a documented advantage for mating with multiple males in the absence of parental care is a kind of snake. Female European adders that mate with more males give birth to more young. This is attributed to increased genetic diversity, because the population studied is isolated and inbred. The more males she mates with, the more likely she would find a distant relative to sire viable offspring. But there could be a more general explanation for why female snakes mate with more males, and why female snakes that mate with more males have more young. But before revealing this idea, it is important to discuss some intriguing aspects of snake courtship and copulation.

Courtship and Copulation

Unlike the "dances" that occur between males for access to females, courtship in snakes is completely horizontal. The male lies on top of the female and tenderly nudges her with his snout near her cloaca and then across her neck. Tongue flicking by the male is rapid and prolonged, occurring much more frequently than typical. Males flick their tongues along the length of the female's back—the tines gently tickling her skin—which is often a fresh surface, oily and musky from her recent shed. He is probably lapping up her pheromones, a behavior akin to inhaling deeply into your lover's dark hair. The male often snaps his head quickly, a behavior termed "head jerking." Sometimes the male's entire body twitches. Males then ripple their bodies down the length of the female, sending waves of coils sliding down her body. This decidedly sexy maneuver has been completely neutered by scientists, who gave it the woe-

A pair of southern Pacific rattlesnakes copulating. *Photograph by Mark Herr*

fully technical term "caudocephalic waves." The female usually sits still during the whole sequence. Occasionally she may begin to crawl away, bringing the courtship to a temporary stop. The male then begins the entire sequence over again from the start, which varies by species but contains similar motifs in all snakes. The female signals her readiness to mate by raising her tail and opening her cloaca. The male then nestles beside her, entwines his tail with hers, and inserts his reproductive organ.

To be precise, he inserts one side of his reproductive organ, a spectacularly hideous contraption known as the hemipenes. The hemipenes is a paired organ normally tucked into the base of his tail. It unfurls slowly—as if turning a sock inside out—and grows like a bunched-up straw wrapper tapped with water. It is truly horrible; it is pink white, and as it unfurls it branches into two main heads, and then these heads can branch again into a pair of long, winding prongs, giving the snake what amounts to a four-headed penis. Snake hemipenes are ornately adorned with despicable spines —often dozens of small spines and a pair of formidable hooked thorns toward the tips. There is a central groove running along the length of each side, which slowly conducts sperm from the male's cloaca into the female. Each snake species has its own uniquely shaped hemipenes, which vary in number of prongs, length, girth, spine shape, and spine density. Nobody is quite sure what in the hell the spines are for, but they almost certainly anchor them inside the female.

The hideous copulatory organ of a Mojave rattlesnake.
Photograph by Noah Fields

The sperm is stored in a duct leading down from the testes, enters the cloaca, travels up the hemipene groove, and is then conducted into both of the female's oviducts by capillary action. Since only one side is used at a time, the organ presumably needs two heads to inject sperm into each separate oviduct. The sperm then travels up the oviducts perhaps half the female's body length to fertilize the eggs. Copulation can last agonizingly long in snakes. A copulation in the cottonmouth was once observed to last eight hours. A copulation in an eastern hog-nosed snake lasted three days. There are observations of females dragging males along behind them by their hemipenes. This may explain why so few snakes have been observed copulating, and almost certainly explains why I've never seen a cottonmouth mating. I highly doubt they do it out in the open.

The reason for the bizarre mating systems of snakes, their spiny hemipenes, and their marathon sex sessions may be related. In many animals, females mate with more than one male, and competition occurs within her body for access to eggs. This is referred to as sperm competition, which has only recently been recognized as a major driving force of evolution. An obvious example is the association between the size of male gonads and promiscuous mating systems.

In species where females mate with many males, males tend to have larger testes, which help them fertilize the eggs amid heavy competition. In species where males are certain of paternity, the testes are often much smaller. A great example of this can be found rather close to home, evolutionarily speaking. Chimpanzees are notoriously promiscuous, with most members of the group mating with all the others. Mountain gorillas have a male-dominance mating system, with the silverback guaranteed to sire most of the young. Chimpanzees have much larger testes relative to mountain gorillas. Humans, a close relative to both species, have a testis size somewhere between these two extremes, and variable mating systems that are also somewhere in between.

Because female snakes mate with more than one male, it is likely that spiny hemipenes allow males prolonged opportunities to fertilize as many eggs as possible. Spinier hemipenes might allow males to anchor for long time periods, perhaps to ensure that subsequent males do not fertilize the eggs. But more generally, the hemipenes may be a rather inefficient way to transfer sperm. Lizards also have hemipenes, but the most elaborate and spiny hemipenes are possessed by snakes. Perhaps prolonged copulation and spiny hemipenes in snakes are simply a solution to the elongated shape of their bodies; in snakes, sperm must travel down the rear of the male, through the cloaca, along the hemipenes, and quite far up the female oviduct to reach the eggs. The longer copulation lasts, the more likely sperm will have time to be properly transferred. This may also explain why females would tend to mate with multiple males. When females mate with one male whose sperm fails to reach all her eggs, she will have reduced reproductive success. By mating with more males, she ensures that all of her eggs will be fertilized and promotes competition among the male's sperm.

The behaviors and hemipene shapes of the red-sided gartersnake and plains gartersnake support this notion. Red-sided gartersnakes mate briefly (15 to 20 minutes, a reasonable time frame) and have simple hemipenes. Male plains gartersnakes attempt to mate for as long as 98 minutes, and females actively struggle to terminate their shenanigans by dragging the males behind them, performing up to 12 barrel rolls per sex session. These males have bilobed, spiny hemipenes. This is a classic example of sexual conflict: for males, the objective is to mate for as long as possible to fertilize as many of her eggs as possible. For females, mating with more males may increase the genetic diversity of her clutch or ensure that all the eggs are fertilized, in case she mates with a dud or a close relative. So females try to disengage males early, and males hang on for dear life.

Female snakes have special crypts within their reproductive ducts that store sperm temporarily or for extended periods of time. All American snakes ovulate at about the same time of year, between April and June, depending on how far south or north the population is. Snakes that mate in the spring can ovulate as much as a month after copulation, which gives sperm from multiple matings ample time to mix and compete before the eggs are fertilized. Snakes that mate in the fall store sperm over the winter and use it to fertilize the eggs the next spring. Prairie rattlesnakes undergo estrous and mating almost a full year before the eggs are fertilized. Sperm can be stored anywhere from 1 to 10 months, depending on the species and the mating season. That is a long time for sperm to compete for the eggs.

Embryonic Development

The development of eggs is different for snakes relative to humans and other mammals. In humans, the egg is ovulated at a microscopic size, becomes fertilized, and then implants on the uterine wall. A connection is formed between the embryo and the mother, and she partitions all the nutrients, energy, and oxygen to the developing embryo until it is born at an incredibly advanced stage of development. In snakes, the eggs are ovulated at a large size, after they have accumulated massive amounts of yolk. Beyond this there are two drastically different developmental patterns seen in American snakes: some lay shelled eggs, and others give birth to live young.

Colubrine, leptotyphlopid, and dipsadine snakes are oviparous (they lay eggs); after yolking, a shell gland in the oviducts secretes a leathery casing around the egg and they are laid after perhaps a month of development. Snake eggs vary in shape from oval to oblong to spherical. They are leathery to papery in texture. Many swell and adhere to one another, forming a large egg mass in which the eggs are stuck together, not unlike a batch of biscuits placed too close together on the pan. The eggs are not waterproof; they have tiny pores that allow gas exchange and absorption of water. Like those of other reptiles, birds, and mammals, snake eggs contain internal membranes that serve various functions like gas exchange, nutrient storage, waste storage, and shock absorption. After they are laid, they grow larger as they absorb moisture from the substrate and as the embryo grows.

Nesting sites vary by species, but eggs are always laid in chambers under cover—in the dirt, sand, or in rotten logs. Rough greensnakes and ratsnakes lay eggs high in trees in rotten hollows. Both species may use the same tree hollow year after year and perhaps use them communally; certain hollows have been discovered with hundreds of hatched eggs. After laying, oviparous snakes abandon their clutch, although exceptions exist. Female mudsnakes and rainbow snakes stay with their eggs until they hatch. Mudsnakes also choose a remarkable location to brood their clutch: several nests with their attending mothers have been found within the nests of American alligators. They can therefore take advantage of the considerable protection provided by a 300-pound adult female alligator, which stays near the nest and viciously defends it.

Nest site selection in wild snakes is poorly understood because nests are difficult to find and eggs are difficult to monitor in nature. Some of our best insight into this topic comes from a meticulous study of northern pinesnake

A hog-nosed snake emerges from its egg.
Photograph by Ashley Tubbs

A ratsnake emerges from its egg.
Photograph by Todd Pierson

A mudsnake female attends her clutch. *Photograph by Ian Deery*

nests in New Jersey. Here, several nests were discovered and their characteristics studied. Nest sites had nice, loose sand and were typically out in the open, so they received good sunlight. Females chose sites where it was easy to dig and avoid roots, excavating elaborate burrows 3 to 12 feet long with nest chambers at the end. Eastern hog-nosed snakes along the shores of Lake Ontario choose similar nesting locations in open dunes with loose sand. Their method is nothing less than astonishing: females dig face-first into the sand while wriggling their bodies, which creates a sort of current of sand fanning out behind them. They then disappear into the burrow, laying their eggs below ground. Their midsection reemerges, their head and tail still in the burrow. With this loop of the body aboveground, they are able to slither sand back into the burrow, filling it and smoothing it out. Additional information about nesting in snakes comes from fortuitous field observations of hatchling snakes and artificially reared clutches. These show that most colubrine eggs develop best at temperatures ranging from 70°F to 86°F, and hatch after an incubation period of 50–85 days, usually in August and September. Dipsadines develop slightly faster, after an incubation period of 47–70 days, and also hatch mostly during August and September.

Rather than laying eggs, boids, viperids, yellow-bellied seasnakes, and natricines give birth to live young. The eggs have thin membranes surrounding them and continue to grow inside the oviducts of females throughout the summer. Pregnant females tend to bask more frequently during gestation, because higher body temperatures decrease development time. Female timber rattlesnakes and prairie rattlesnakes congregate in gestation "rookeries"—rock outcrops that receive abundant sunlight—for this purpose. Pregnant cottonmouths bask on top of stumps, hummocks, or levees providing

Young copperheads are born alive within a thin membranous sack. *Photograph by Patrick Thompson*

sunny swamp openings. Pregnant watersnakes often bask stretched out along branches over the water's edge, allowing them a place to bask with a quick getaway if an enemy approaches.

Some snakes even have a placental connection with their young and provide additional nutrients and gas exchange beyond the yolk. Gartersnakes are the best known of our placental snakes, although the connection between the young and mother is not as intimate as it is in mammals. Cottonmouths do not have such a connection, but there is exchange of gasses between the mother and young. In boids, pitvipers, seasnakes, and natricines, the egg is simply retained by the female and laid as a membranous sack. The young then immediately break free from the sack and are born. Natricine snakes typically give birth around July through September after a 2.5- to 3.5-month incubation period. Pitvipers usually give birth after a slightly longer 4-month incubation period, usually during August and September.

Life History

The energy invested in developing eggs and embryos has been the topic of many studies and is important for determining the long-term population trends in snakes. The amount of investment that animals devote to their young over their lifetimes is an important component of what is known as their life history, and this is quite variable in American snakes and critical to our understanding of their ecology and conservation.

Most of the energy required to make babies comes from massive fat reserves stored in the snake's body. Snakes pack away these fat reserves after feeding and only reproduce when they have enough to produce a clutch of eggs or a litter of newborns. Some snakes are able to supplement the energy they package into growing babies by feeding while yolking the eggs. Snakes that are experimentally fed in the laboratory or in the wild are able to produce bigger clutches, reproduce more frequently, and transfer nutrients from their recent meals directly to the eggs. The problem for most snakes is that it is difficult to feed with a bellyful of eggs or developing embryos. Egg layers avoid this problem by getting rid of the eggs as soon as possible. Live bearers have smaller clutches, and many species don't feed while gestating. This is either because of lack of room in their body cavity for both food and eggs, or because behaviors that support gestation (basking) are not compatible with feeding. Some snakes find a way to continue feeding while pregnant (western diamondback rattlesnakes, northern watersnakes, and black swampsnakes), and can increase their reproductive output. The crucial factor that explains most of the diversity of reproductive productivity among American snakes is feeding success. Females that feed more often grow larger and have bigger clutches more frequently. Following bad years with low prey numbers, snakes reproduce less.

There is considerable variation in how frequently snakes reproduce, both among species and among individuals within the same species. The eggs begin

Emily Taylor

Unraveling the Sex Lives of Rattlesnakes

Of all the snake biologists interviewed for this book, Emily followed the least likely path to becoming a scientist. She was more into sports and was an athlete growing up. Her family never did outdoorsy things, and she was "not particularly into nature and wildlife." She was an English major at Berkeley when Emily ended up in the Natural History of the Vertebrates class taught by Harry Greene. The course had field trips where students saw as many as 40 amphibian and reptile species. On the first field trip to Mendocino County, they uncovered a glossy black-and-white ringed snake—the gentle and handsome California kingsnake. Then, according to Emily, "Everything went quiet. The edges around my vision went dark. It was the coolest thing I'd ever seen in my life."

Her destiny irrevocably changed, Emily took a new academic path and studied western diamondbacks in Arizona for her PhD research. She learned from her mentors the benefits of hypothesis-driven, experimental research. She loves this approach so much she told me that she "dreams about it at night." I don't doubt it: her field experiments with rattlesnakes show that females can produce many more young if supplemental food is available; this finding expanded upon observations that rattlers reproduce much more in good, or rainy, years. Recently, she examined relocated northern Pacific rattlesnakes and showed that the brain region used for spatial navigation became larger in the snakes that had been displaced the farthest. Her research applies powerful experimental approaches to field problems, which makes her studies much more valuable.

Chris Leschinsky

Emily is now a professor of biology at California Polytechnic Institute in San Luis Obisbo. San Luis Obisbo fits the dictionary definition of "quaint California town." It has spacious streets, beautiful people, varnished Western-style saloons, and outdoor dining on 73-degree afternoons in June. I recently had the chance to meet up with Emily and follow some of her students around as they tracked their rattlesnakes. Her students are always bright, original people, fun and enthusiastic, making me wonder if she's rubbed off on them or whether like attracts like. They're always the life of the party at our national herpetology conferences, and I'm slightly jealous of where she lives and how good she is at her job. She says, "It's a joke how much I love my job . . . this is the best job in the world." She will be turning out key figures in our field for years to come, and I like to imagine her there, sipping wine in that beautiful, sunny country.

Emily and I shared a graduate advisor when I was just getting started and she was finishing her PhD. She's my academic big sister. Like my real big sister, I've always looked up to her, and I've been impressed by her work. I remember her easily tackling needling questions from an antagonist at a conference, and a few years later leading a seminar on snake reproductive biology. When I was starting my PhD, she got her faculty job at Cal Poly. Now I've got a faculty job, and she's head of her graduate program. She's always been there, one step ahead, leading the way.

development around the same time so that the breeding season is constant from one year to the next. If a snake misses this window, she will skip mating for a year or more. When she finally stores enough energy to reproduce, it will be on schedule with the rest of the females in the population. Females of some species (eastern wormsnakes, ringneck snakes, red-bellied snakes, western ribbonsnakes, northern watersnakes, and brown watersnakes) can produce a clutch every single year, which means that each year all mature females in the population reproduce. In most other species, including egg layers (glossy snakes, western wormsnakes, racers, rough greensnakes) and live bearers (probably all pitvipers), some proportion of females fail to reproduce every year. Each year some females produce young, while others mate in anticipation of reproducing the following year. Females with meager fat reserves may not reproduce for several years in a row.

The number of young produced by snakes depends on the size of the snake, her energy stores, and genetic controls. Females devote anywhere from 10% to 60% of their body weight to developing young. Average clutch size varies greatly in American snakes, from a low of just 3 eggs (western wormsnake, ringneck snake) to as many as 28 (brown watersnake). Sometimes snakes produce only one baby at a time, however an impressive record of 111 eggs laid by a single mudsnake gives you an idea of how many young that large females can produce when in their prime. Snakes have the capacity to produce either a large number of small young or a smaller number of large young, and this seesaw of possibilities is controlled genetically. By artificially feeding snakes, you can get more eggs, but their size at hatching does not change.

A final component of life history that can influence the reproductive output of a snake population is time to maturity. Some snakes quickly become sexually mature, perhaps within a single year of hatching, while other snakes may delay maturity for several years. This is probably determined genetically and also by food supply; even in snakes that are geared toward slow maturity, females likely become mature earlier when they acquire energy more quickly.

Clutch size and reproductive frequency contribute to the characteristics and persistence of snake populations. Some snakes are able to mature quickly, reproduce frequently, and produce large numbers of small young; such snakes may be more successful in unpredictable habitats and therefore share life history traits with other "weedy" species. Others have delayed maturity in order to produce large young only occasionally; such species often adopt this type of life history strategy in more predictable and mature environments. But certain life history strategies now conspire to make some snakes especially vulnerable to human exploitation.

This is perhaps best illustrated by comparing life history characteristics of three kinds of rattlesnakes, all of which endure an incredible level of attrition at the hands of humans. All rattlesnakes are killed by the thousands every year by people and their automobiles. But some rattlesnakes are also systematically exterminated during annual "rattlesnake roundups" hosted by small towns throughout America. Once much more common, these roundups have thankfully declined in the past couple of decades, mostly owing to the difficulty in finding snakes to round up. Still, roundups remain popular events that generate an economic boost for their otherwise anonymous and grim towns—towns like Sweetwater, Texas; Opp, Alabama; Noxen, Pennsylvania; and Wigham, Georgia. Roundups have been harvesting and killing large numbers of rattlesnakes for decades, and according to their own records, they are annihilating local populations of rattlesnakes. The most common snakes rounded up are timber rattlesnakes, eastern diamondback rattlesnakes, and western diamondback rattlesnakes. Despite the heavy toll taken by the roundups, however, only two of these are declining and in danger of extinction.

Timber rattlesnakes have been cleared out of the Appalachians and northeastern United States. In such areas, the growing season is short, so timber rattlesnakes take as long as 10 years to reach maturity, and after that they usually take 3 or sometimes even 4 years to produce their next litter of just seven babies. Because reproductive females are the easiest to locate (they bask out in the open in rookeries), rattlesnake catchers quickly decimated populations of these snakes. Eastern diamondback rattlesnakes delay maturity for as long as 7 years, have relatively small litters, and take anywhere from 2 to 4 years to produce a new litter. This magnificent rattlesnake has been severely affected by habitat loss and direct killing due to the rattlesnake roundups. If it was not a rattlesnake, it would probably already be protected as an endangered species. Both timber and diamondback rattlesnakes have life histories that cannot withstand heavy exploitation.

The biggest, baddest, nastiest rattlesnake roundup of them all (the Texas-sized Sweetwater roundup) kills mostly western diamondback rattlesnakes, but there have apparently been no declines in the numbers of snakes they gather every year. And there are still plenty of western diamondback rattlesnakes in west Texas. Nowhere in the United States is this snake considered a conservation concern. The western diamondback rattlesnake, which is similar in size to both of the rarer species mentioned above, is much more of a habitat generalist and thrives in brushy mesquite country, oak-juniper woodlands, succulent bajadas, and low barren deserts from Texas to California. And their innate life history characteristics also give western diamondbacks a distinct edge in the face of human persecution. Western diamondbacks are among the few pitvipers known to have a high proportion of reproductive females every year (up to 68%). They can do this because they are good at hunting and rapidly build up fat stores. They are especially adept at reproducing after wet winters in the Sonoran Desert. Western diamondbacks also continue to feed throughout their ovarian cycle, all the way through pregnancy, which is unusual for pitvipers. Supplemental feeding during pregnancy allows them to produce clutches more often, and female western diamondbacks are in much better physical condition after giving birth than most snakes. They have large litters for their body size, and they become sexually mature in as few as three years. Compared to the two imperiled eastern rattlers, western diamondbacks breed like rabbits, which enables them to maintain steady populations even in the face of roundups.

Maternal Care

After they hatch, young snakes probably never see their parents again and have to make their own way in the harsh world. Because many reptiles are oviparous and do not provide parental care, they are often thought of as heartless compared to the good parents known among birds and mammals. But some exceptions are known and provide us with unparalleled insights into how parental care developed in the first place. Alligators dutifully attend their nests and assist their hatchlings out of the eggs, responding to their cute little clucks by gently cracking the eggs with their monstrous jaws. Baby alligators can then be observed clamoring on their mother's back and following her around for up to three years, almost like scaly ducklings. Pythons wrap themselves around their eggs and provide warmth for development by shivering their formidable constricting muscles. King cobras also attend their clutches and provide con-

A female Arizona black rattlesnake and her young. *Photograph by Timothy A. Cota*

siderable defense against any animal foolish enough to try to steal an egg. Egg attendance is fairly common among squamates (in the United States, skinks, glass lizards, and alligator lizards attend their eggs until hatching), although in American snakes it is rather exceptional (mudsnakes and rainbow snakes are the only two species known to do it). But perhaps the most astonishing form of parental care among reptiles is exhibited by pitvipers.

After they are born, rather than immediately leaving the vicinity of their mother, baby rattlesnakes, copperheads, and cottonmouths stay with their mothers for about a week. During this time they stay piled around, on top of, or underneath the coils of their mother. She doesn't provide food, but there are reports of pitvipers aggressively defending their young during this time. Prairie rattlesnakes are more likely to rattle furiously just after giving birth. Prior to giving birth, they are more likely to lie still and rely on their camouflage. If the mother is removed from her young, she will crawl back to them. The maternal attendance period is associated with shedding; the first shed of the young is within about 7 to 10 days of birth. Immediately prior to shedding, snakes are more vulnerable because their clouded eye scale obscures their vision. After their first shed, either the mother crawls away or the young crawl away from the mother. So, it appears this brief attendance period is for the protection of the young during this vulnerable time.

To understand the development of these behaviors, the best we can do is to compare many unrelated species and their behaviors and concoct an evolutionary "scenario": a story about the steps that were perhaps involved in the development of the behavior. Parental care is often thought of as a trajectory, starting with bad parents who immediately abandon their eggs (turtles, many lizards, many snakes) and ending with the remarkably complex, monogamous parental care seen in some birds and mammals. Between these extremes lie hints at what the intermediate steps may have been: alligators attending nests of rotting vegetation and helping the young hatch; Australian birds building similar mound nests, compulsively regulating the temperature within the mounds, then providing no care beyond hatching; birds like ducks, whose young hatch nearly fully formed and are provided with limited care by the mother for just a short period; pigeons who are born eyeless, puny, and featherless, and are provided extra nutrition from the mother with a gooey milk secreted by her crop; marsupials born as blind, hairless gummy bears who must find a teat and are then cared for within a pouch; humans born helpless and provided everything until they graduate from high school, and in some cases, college.

By investigating the most primitive, most incipient form of parental care known, we may determine the initial steps involved in the development of what has until recently been thought to be patently mammalian behavior. And this might best be illustrated by the dreaded venomous snakes within our American landscape, who are in fact good mothers.

6 · Snake Food

Several species of snakes . . . have been found on being killed, to have a Squirrel in their stomach; and the fact that Squirrels, birds etc., although possessing great activity and agility, constitute a portion of the food of these reptiles, being well established, the manner in which the sluggish serpent catches animals so far exceeding him in speed, and some of them endowed with the power of rising from the earth and skimming away with a few flaps of their wings, has been the subject of much speculation.

JOHN JAMES AUDUBON, *The Quadrupeds of North America*

A common gartersnake on the prowl. *Photograph by Curtis Callaway*

Urban Jungle

The bluffs overlooking the roiling, charging Ohio River were dressed with new green, the sandstone outcrops mud yellow and covered with vines. Dozens of lizards ostentatiously warmed themselves on rocks, gearing up for the summer months and occasionally darting out to snap up a fly. At the base of an old crumbling homestead, a plume of termites emerged and floated on the soft spring wind, their wings catching the morning light. Sparrows dipped down from a pink-blossomed redbud, snapping up as many of the insects as they could. My wife and I looked for snakes, hoping to find a Kirtland's snake— a colorful snake peculiar to the Midwest with even more peculiar habitat tendencies. We began finding dozens of gartersnakes, as many as five under a single board. Some had already emerged and were sunning themselves in the grass along the edge of old rock walls. I captured one that had an obvious bulge in its belly—it had fed recently. Even though I was certain this was the gartersnake's first meal of the year, my curiosity eclipsed my guilt, and I gently squeezed the snake just below the bulge, working it back up toward the head like a tube of toothpaste. This method, called "palpation," or simply "palping," is an easy way to determine the diet of snakes. The snake opened its mouth and regurgitated a half-digested lizard, of the same species that we saw by the dozen sunning on rocks.

It turned out this lizard had never been identified as a prey item for the common gartersnake, despite the snake having the widest distribution of any American species. It lives from Florida to Maine and beyond, west to California and Washington, and therefore its range overlaps with nearly every kind of potential snake food in the country. Gartersnakes also have one of the most catholic diets of any American snake: they are known to feed on several kinds of invertebrates like earthworms and leeches, as well as a variety of vertebrates ranging from frogs and salamanders to rodents and small birds. And it got better. We realized that, somehow, not only had this specific kind of lizard never been reported in the diet of gartersnakes, lizards in general had, until that May morning, never been recorded as a meal for this common snake. So we busily documented the observation so we could publish it as a brief natural history note in our favorite scientific periodical, *Herpetological Review*. We measured the snake and the lizard, got pictures, and obtained a precise locality using my handheld GPS unit. The location of this wild scene of snakes killing lizards and a feeding frenzy of sparrows was 39.11919° N latitude, 84.51722° W longitude—otherwise known as Cincinnati, Ohio.

Sometimes you find snakes where you least expect them, and you don't have to travel to the wilds of Brazil or Borneo to do science. In fact, the peculiar habitat tendencies of the Kirtland's snake means that the best places to find them are old vacant lots in cities like Cleveland, Cincinnati, and Indianapolis. A colleague told us to look along Vine Street in Cincinnati, where he said they were common along the rock walls leading up to Mt. Auburn, the bluff overlooking the Ohio River where President Taft grew up. There too you can find dozens of common wall lizards, a pet store escape from Europe that has become established and common in the city. Traces of the bluff's once wild aspect were visible but otherwise replaced by urbanization. The area is blocks from downtown and looked like the set of David Simon's *The Wire*: vacant brick row houses and adult men hanging out on stoops during working hours on a weekday, occasionally ducking briefly into passing cars. The folks were nice though, most smiling as they walked past, and some even asking what we were up to and guessing we were there to find snakes. One guy called down from a third-story window, and we spoke for a few minutes before he finally asked what must have been to him a burning question: "You're rich, right?" I suppose he must've thought that all scientists are rich, or all folks who look remotely waspish. In reply, I simply pointed over to my beat-up 1997 Ford

Ranger, whose mileage turned 300,000 that summer, and said, "Have you seen my truck?"

Although we failed to find the rare, city-dwelling Kirtland's snake, we did end up contributing to our understanding of snake diets. Before stopping off along Vine Street to look for Kirtland's snakes, 95 vertebrates were known in the gartersnake's diet. After we left, the list rose to 96, and common wall lizards were confirmed as prey in gartersnakes for the first time.

Because of their tiny metabolism, snakes do not feed frequently. Large, infrequent meals can sustain a snake for a long time, and some snakes might only feed 12 to 15 times a year to satisfy all their needs. A large meal might take a couple of days or a week to become completely liquefied in the stomach and move into the intestine. So, if you see a bulge in the stomach, it indicates a recent meal. Most snakes you find exhibit no such bulge.

Much of what we know about the diet of some snakes, especially for rare species, comes from chance observations just like our discovery in Cincinnati. Snake biologists occasionally luck out and find a snake with food in its belly, coax it out, and report these single observations. More information comes from repeated capture and palpation of many individuals from locations where a species is common. Therefore we know a great deal about the diet of gartersnakes throughout their range, and enough studies have been published to allow geographical comparisons for some snakes. Finally, a great deal can be learned about the prey of snakes from dissections of preserved snakes present in museum collections throughout the country. Often only about 15% of museum specimens contain identifiable prey items, so researchers must examine hundreds of specimens to find useful information. Taken together, however, chance observations, detailed field studies, and museum studies reveal a great deal about the hunting tactics and preferred prey of American snakes.

Killer Tactics

The tendency for most people is to sympathize with mice killed by snakes and elk chased mercilessly by wolves. And it is difficult not to feel sympathy for a frog being slowly swallowed alive by a snake, squealing with terror. They cling desperately to a blade of grass as they disappear down the snake's throat, and continue squealing from the belly of the snake. It is difficult not to feel sympathy for a big, brown-eyed chipmunk struck by a timber rattlesnake, gasping for its last breath, bleeding from twin puncture marks. Still, I am a great fan of predators. They are important regulators of prey populations, and countless examples of landscapes sanitized of predators demonstrate the impact of herbivores run amok. In natural systems it is often difficult to know the true impact of the predators, but unbalanced ecosystems—either those without predators or those with exotic predators—offer a powerful lesson. Predators have harder lives than prey species, and, because it is much easier to be chased than to give chase, they must be far better at what they do than their prey.

A banded watersnake swallows a southern leopard frog alive. *Photograph by Pierson Hill*

Starvation is the fate of many snakes. And many snakes are both predator and prey, so they experience the worst of both worlds.

All snakes are hunters, and none are vegetarians or even omnivores. As we have seen previously, the daily hunting strategies of snakes range from ambush-hunting pitvipers to raiding ratsnakes and pursuit-hunting coachwhips. But none of these snakes establish hunting grounds randomly and expect prey to simply show up. Snakes use their exquisite sense of the tongue and vomeronasal system to increase their chances of finding food, and many have genetically programmed preferences for the scent of their preferred prey. An ingenious experiment showed that pygmy rattlesnakes will quickly show up along trails laced with frog scent. Similarly, snakes that probe in and out of rodent burrow systems can probably detect tunnels frequented by rodents versus stale old burrows without food. Pursuit predators are our only snakes that favor visual cues for their hunts, and their exceptionally large eyes help them track and chase prey.

A demonstration of the power of the tongue follows the strike of venomous pitvipers. The snake briefly latches on to the prey, usually just at the shoulder to inject venom into the chest cavity of a mouse. Then it lets go to avoid the gnawing teeth of the enraged rodent. The rodent will scamper away several meters before succumbing to the potent venom. Like snipers, snakes lie still and concealed for some time after striking, as if to avoid retaliation. After a few minutes, the snake begins a rapid session of tongue flicking, referred to by the admittedly tedious acronym SICS, which stands for strike-induced chemosensory searching. The tongue flicks rapidly, and the two sides of the tongue tips allow the snake to directionally sample the environment and follow the trail of their prey. Snakes can even differentiate a mouse they have bitten versus one experimentally placed; along with the potent tissue-destroying components and paralyzing factors, the venom carries some sort of tracer. It can take a rattlesnake anywhere from 15 minutes to several hours to successfully track down their prey, and by this time the venom has killed the mouse and also accelerated digestion.

Some prey are captured and held by venomous species instead of being released and tracked. Cottonmouths strike and hold on to frogs and salamanders, and rattlesnakes strike and hold birds. This is probably because frogs and birds are harder to track after being released, and neither are as dangerous to handle as a gnashing rodent. There are several venomous species that seize and chew venom into their prey before swallowing them. These include our dangerously venomous coralsnakes, as well as the mildly venomous, harmless species like ringneck snakes, black-headed snakes, crowned snakes, lyresnakes, cat-eyed snakes, and nightsnakes. For some of these species, the effects of the venom have been confirmed; although harmless to humans, it quickly subdues their prey.

Nonvenomous species do not have the use of such a streamlined and effective killing apparatus. Many small snakes cannot handle large or dangerous prey and simply feed on tiny, soft-bodied creatures like worms, salamanders, and frogs. These snakes feed in perhaps the most heartless way possible: many

A western diamondback rattlesnake swallows a mourning dove. *Photograph by Bob Herrmann*

A ringneck snake swallows a dusky salamander alive. *Photograph by Todd Pierson*

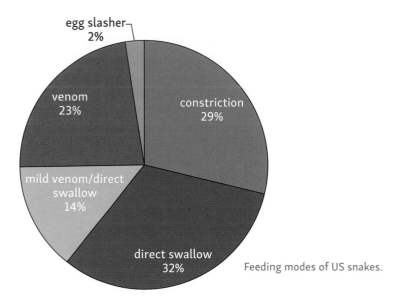

Feeding modes of US snakes.

American snakes simply seize their prey and swallow them whole while they are decidedly alive. The "direct swallow" feeding mode is also employed by some larger snakes that feed on more formidable items. But even indigo snakes must chew the heads of rattlers until they are dead before eating them.

Constriction is a common feeding mode in our larger harmless snakes that include mammals in their diet. Constriction involves wrapping several loops of the body around the prey and using the powerful trunk muscles to squeeze. It is a dangerous strategy that brings snakes into close contact with vicious prey that are desperately trying to defend themselves. But it is also quite effective; rats can be subdued in less than a minute; compression of the chest cavity

A Great Plains ratsnake constricts a cactus wren before swallowing it. *Photograph by Noah Fields*

prevents the heart from beating and induces cardiac arrest. Snakes feel the heartbeat of their prey through their skin, and cease constriction when the heart stops. Ectothermic prey like lizards and snakes are constricted as well, but since they have such a low metabolic rate, it often takes much longer to kill them. Sometimes they aren't even killed, and instead they are simply weakened enough to permit swallowing. Nearly one-third of American snakes use this frightening but effective killing mode, including both of our boids and a group of colubrids with a mostly American distribution: the milksnakes, ratsnakes, gophersnakes, and their relatives. Several burrow-raiding species use a variant of constriction in the tight spaces within rodent burrow systems. In these tunnels they cannot completely wrap around their prey, so they instead pin their prey to the walls of the burrow with a single coil to immobilize them for swallowing. With few exceptions, snakes do not chew or dismember their prey and ultimately swallow their prey whole.

Although their methods might appear simultaneously sophisticated and cruel, like all predators, snakes are usually unsuccessful in their hunts. Timber rattlesnakes establish ambush sites for as little as two hours to nearly three days, and encounter about one rodent a day. Of these, they only strike about 75% of time, mostly because the rodents scamper by so fast that the snake has no time to react. Eighty-seven percent of strikes miss. After the strike, snakes may take several minutes to a few days to then successfully track down and consume their food, and sometimes they fail to find it. These are not exceptionally good numbers, but for most predators this is the sad reality.

Generalists and Specialists

In 2001, Scott Boback, Roger Birkhead, and Matt Williams began studying cottonmouths in the swamps of Tuskegee National Forest in Alabama. They waded across a beaver marsh once a week, catching as many snakes as they could, and transported them 20 miles back to the lab for processing. To safely transport dozens of cottonmouths, Roger invented the "cottonmouth condo," a portable carrier with eight individual tubes that could each house a venomous snake. It looked something like the barrels of a Gatling gun and was probably ten times as dangerous to carry around. The little caps at the ends of the tube were always getting rubbed off in the thick swamp brush, and the guys sometimes made it all the way back to the car before noticing a smiling cottonmouth dangling from the condo. I ended up taking over their study after they moved on, and I inherited a condo. When I used it, I mummified the snakes in the condo with a heavy layer of duct tape.

Once the guys caught a cottonmouth with an outrageous bulge in its stomach. It was so distended and misshapen it couldn't fit in the condo, so they transported it back to the lab in a pillowcase. They decided to have the snake x-rayed at Auburn's world-class veterinary school. The image showed clearly that the cottonmouth had consumed a large bird's wing. It was just the wing. Still puzzled, they consulted one of Matt's old professors at the University of Florida who was an expert in bird anatomy. From the x-ray he was actually able to identify the wing's owner. Using considerable technology, expertise, and pluck, they were able to add a blue-winged teal's wing to the known diet of the cottonmouth.

Cottonmouths, more than any other American snake, are opportunists that feed on anything that comes within reach of their fatal jaws. Over 170 kinds of vertebrates and many kinds of invertebrates have fallen prey to this king of the southern swamps. The list of species it has consumed includes nearly all the venomous snakes it coexists with, plus dozens of fish, frogs, lizards, rodents,

Rulon Clark

The Watcher

Rulon grew up outside of Provo, Utah. Even though his parents are Mormons, according to Rulon, they had an "academic bent, but not toward science." And they made the fatal mistake of religious fundamentalists everywhere: in their zeal to shelter Rulon from the vices of society, they prohibited him from watching television and made him play outside instead. My Catholic parents did the same thing. If they had only let us watch television, by now we would surely be normal religious folks just like everyone else. Instead, Rulon became absorbed with the lizards and praying mantises near his home. He's not a Mormon anymore. Nor am I Catholic.

Jeff Lemm

Rulon Clark became a biologist and contributed a tremendous amount to our understanding of the secret lives of snakes. These discoveries came from using remote cameras to observe the hunts of timber rattlesnakes. People had examined rattlers hunting before using telemetry, and Rulon wondered, "Why aren't there people out there taking behavioral data?" He soon learned that to observe an entire ambush session is beyond the human attention span. It is simply incredibly boring, and can go on for days. Armed with cameras, Rulon began filming their hunts and watching them on fast forward. By reviewing days and days of video and noting when something finally happened, Rulon has shown how rattlers hunt, how frequently they fail, and how often they give up and move to a new location. I asked him whether he thought snakes are conscious when they are waiting in ambush. I've watched cottonmouths for hours and have come to the conclusion that they are asleep until potential prey approaches. He told me that was pretty much his impression too; they seem to become "startled awake" when prey approaches. Sometimes rodents run by so fast the snake never rouses. He agreed that "sleeping snipers" was a suitable description of their tactics.

Now Rulon has taken his cameras and other technologies to the West Coast, where as a professor at San Diego State University he studies the interactions between rodents and rattlesnakes. He's obtained remarkable footage of squirrels discovering and then harassing snakes, and discovered that kangaroo rats use subsonic thumping sounds with their feet to antagonize sidewinders. His videos and research offer unparalleled insights into the everyday lives of snakes, revealing the amazing interactions between these predators and their prey. His best videos show successful strikes: the rodents never discover the snake, and they receive a fatal bite before dying in agony.

Depriving Rulon the ordinary pleasures of Saturday morning cartoons resulted in his ascendance in science. And now he spends hours and hours watching violent TV.

Cottonmouths are known to congregate at drying water bodies to consume fish trapped in the shallows. *Photograph by Pierson Hill*

and birds. Although its scientific name *piscivorus* means "fish eater," it is by no means reliant on fish as prey. Fish and frogs are typical fare, but each make up only about 30% of the diet of cottonmouths, the rest being filled out by various other creatures. Their common hunting strategy is to wait at the water's edge for prey to come along. The shore of southern swamps is a busy highway used by many animals: here cottonmouths bushwhack fish and aquatic salamanders that feed in the shallows; snatch up frogs that prowl the edge; seize snakes that patrol the edge, looking for frogs; and ambush birds and rodents that visit the edge for a drink. But there are even more curious hunting methods: cottonmouths on Seahorse Key off the Gulf Coast of Florida wait under rookeries for fish dropped by herons. One cottonmouth was observed crawling along a drift fence and lowering its body into the buckets, helping itself to the frogs being studied. It is one of the few snakes in America that frequently eats carrion, explaining how the Tuskegee cottonmouth ended up eating the blue wing of a teal. Because a vast majority of snakes do not dismember their prey, the cottonmouth must have come upon a duck carcass after it was fed upon by a pickier predator that discarded the bony wings.

Animals with few gustatory scruples are known as dietary generalists. Cottonmouths, common gartersnakes, common watersnakes, ratsnakes, and racers are each known to have fed upon over 90 vertebrate species. But even these accomplished killers favor certain prey. For example, insects—especially grasshoppers—are the most common component of the diet of racers in several regions of the United States, despite their having taken over 100 kinds of vertebrates. Gartersnakes are known to have consumed 96 kinds of vertebrates, yet in most places they subsist mostly on amphibians or earthworms. Common watersnakes feed mostly on fish, but some populations are fonder of frogs. So being a generalist isn't just about the variety of prey consumed, but should rather be considered some combination of the variety of prey types and their even spread across the diet. A few more snakes have generalized diets despite having fewer known prey items; indigo snakes and wandering gartersnakes have each consumed far fewer known prey items than cottonmouths and other generalists, but they have no clear feeding preferences.

On the other end of the spectrum are snakes with specialized feeding habits, which for the most part feed on a single type of prey. The most specialized snake diet in the United States is that of the short-tailed kingsnake, which feeds only on another kind of snake: the Florida crowned snake. Such narrow diets are extremely rare in nature, but for obscure reasons many snakes have a peculiar tendency to specialize.

In the United States there are 12 snakes that feed mostly on elongate, soft-bodied invertebrates like earthworms, leeches, and slugs. These include the brownsnake, red-bellied snake, lined snake, earthsnake, sharp-tailed snake, four kinds of gartersnakes, and two kinds of wormsnakes. These snakes have innate preferences for the scent of their prey and are able to swallow worms despite heavy secretion of slimy mucous. Many have the capacity to groom themselves after feeding by rubbing their snouts on abrasive sub-

A red-bellied snake approaches a snail. *Photograph by Curtis Callaway*

strates. Dekay's brownsnakes and red-bellied snakes use another trick to feed on snails: they have long, thin teeth that enable a good grip, and both species slowly and carefully approach snails while they are still out of their shells. They then dart forward to seize the soft body and perform a series of barrel rolls to twist them out of their shell.

Elongate invertebrates like centipedes and beetle grubs are the food of eight species of crowned and black-headed snakes. Centipedes can be fearsome, venomous enemies and can inflict a painful if not fatal bite. Black-headed snakes are able to feed on unbelievably large centipedes (the centipede *Scolopendra heros* can be a foot long) without any evident trouble. These snakes are harmless to man but have elongate teeth in the rear of their mouths and immobilize their prey by seizing them and chewing venom into their body.

Insects and spiders are the mainstay of the two kinds of greensnakes found mostly in the eastern United States, and in most studies, 90% of the diet of racers is also made up of insects, especially grasshoppers. The desert Southwest is home to another series of snakes that feed mostly on arthropods, including formidable prey such as scorpions. These include mostly nocturnal, burrowing species like groundsnakes, sandsnakes, shovel-nosed snakes, and hooknosed snakes. Some of these are easily able to handle their creepy prey and simply swallow them whole. The Mojave shovel-nosed snake seizes scorpions by the tail, which enables them to pin down their body with a coil prior to

A Trans-Pecos black-headed snake seizes its formidable prey: a large centipede. *Photography by Zack West*

swallowing. Hook-nosed snakes have elongate rear fangs and perhaps venom that rapidly immobilizes the spiders, scorpions, and centipedes it eats. The threadsnakes of the Southwest also eat many kinds of arthropods, but a large part of their diet consists of termites and ants. Threadsnakes are capable of living inside the colonies of social insects without being recognized as a threat; they secrete a chemical cloaking signal that mimics the chemical signals of ants and renders them invisible to their prey. Brahminy blindsnakes and Texas threadsnakes are unusual among snakes in their habit of dismembering prey before eating them; they handle termite soldiers roughly and smear them against the edge of tunnels to smash their well-armed heads off before consuming their soft bodies.

North America, and the southern United States in particular, is the center of world crayfish diversity. Nearly every wetland and stream type—from seasonally dry wetlands to blackwater swamps, from isolated limestone springs to giant alluvial rivers—has its own distinctive crayfish fauna. Over 350 species are known in North America. This abundance of prey has not gone unexploited by snakes. Four species are specialized feeders on crayfish. The queensnake and Graham's crayfish snake have taken this specialization even further and feed mostly or solely on crayfish that have recently molted; after they shed their exoskeleton, they are soft and vulnerable and easy to swallow whole. After their shell hardens—as every kid who grew up catching craw-dads knows—crayfish are sturdy and ornery and capable of delivering a spirited pinch with their front claws. Striped and glossy swampsnakes handle hard crayfish with their coils, and the glossy crayfish snake is capable of actively constricting them to death.

All told, about a quarter of our 159 snakes feed mostly on invertebrates ranging from termites to crayfish. Many of these occasionally include small numbers of vertebrates in their diet, especially as they grow larger. The rest of our snakes feed mostly on vertebrates, although they too occasionally feed upon invertebrates, especially when they are young. Many snakes show distinct changes in dietary preference with age, and some even have inherent prey-scent preferences that change as they reach adulthood. For example, striped swampsnakes begin life eating the aquatic larvae of dragonflies, but adults

A queensnake feeding upon a recently molted crayfish. *Photograph by Tiffany Wilkins*

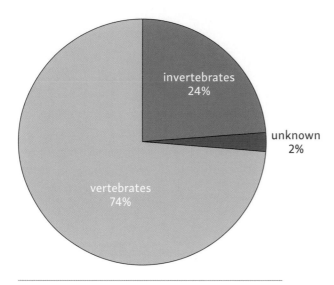

General feeding habits of US snakes.

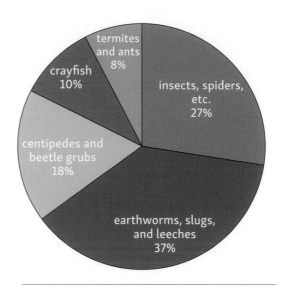

Diet preferences of US snakes that feed mostly upon invertebrates.

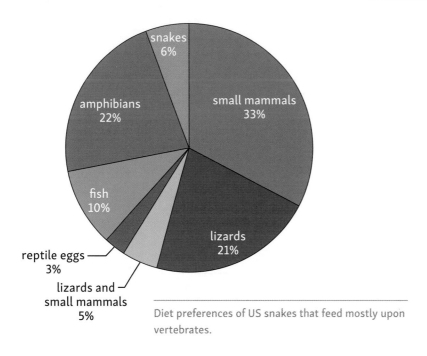

Diet preferences of US snakes that feed mostly upon vertebrates.

switch to crayfish. Young copperheads eat more invertebrates than adults, which are more likely to consume small mammals. Several rattlesnakes (ridge-nosed rattlers, southern Pacific rattlers, and speckled rattlers) begin life eating mostly lizards or centipedes, then graduate to rodents at adult size. There is also a peculiar tendency for many snakes to feed on larger and larger prey as they grow, and these larger snakes will even pass up smaller prey if given the opportunity to feed on them. Other snakes continue feeding on small prey as they grow, in addition to the large prey they are only able to handle at larger sizes.

Fish are the dominant prey of 11 of our snakes, including the venomous seasnake, eight species of watersnakes (genus *Nerodia*), and a strange garter-snake of the southwestern mountains. The glossy and gorgeous rainbow snake of southern spring-fed streams is extremely specialized and feeds mostly on American eels. Many kinds of fish, including spiny kinds that seem impossible to swallow, make up anywhere from 60% to 98% of the diet of several kinds of watersnakes. Brown and diamondback watersnakes specialize further by preferring catfish. The heavy spines of catfish represent a serious obstacle for a snake, which does not have fingers for manipulating prey. Brown watersnakes

A rainbow snake feeding upon an American eel. *Photograph by Rose Williams*

A Florida green watersnake swallowing a catfish. *Photograph by J. D. Willson*

bite catfish and slowly work their way up to the spines. They wait until the catfish flails and unlocks the spines, then quickly push their jaws down to flatten them. They can fully swallow a catfish in this manner in only 15 minutes. Watersnakes are sometimes found with the spines protruding clear out of their bodies, but they are somehow able to digest and heal around the punctures. Watersnakes hunt in a variety of ways, including deliberate, steady underwater patrolling; sweeping their open mouths in sideways arcs in the shallows of streams and ponds; and lying in wait along the water's edge, snatching passing fish by ambush. Some swim along until a fish swims near their side, then quickly circle back on themselves, trapping their prey inside the loop. The most remarkable strategy is employed by the mangrove saltmarsh snake of the Everglades, which lies coiled among the stilts of mangroves and deliberately flicks out its tongue in a loop, tickling the water's surface. It does this slowly and patiently, eventually attracting small fish. When they come close enough, it strikes. Tongue luring is also practiced by the completely unrelated aquatic gartersnake of California.

Nineteen American snakes feed mostly upon amphibians, especially frogs and tadpoles. These include some rare snakes of the subtropical region of Texas like the speckled racer, regal black-striped snake, and cat-eyed snake, the latter of which is even known to feed on frog eggs. Plain-bellied and banded watersnakes prefer frogs to fish, and eight kinds of garter- and ribbonsnakes feed mostly upon frogs, tadpoles, and salamanders. Further specialization can be found among these amphibian eaters. Toads, which are toxic to most predators, are relished by hog-nosed snakes. Hog-nosed snakes tolerate the poison and use their enlarged rear fangs to pop toads, which try to thwart their

A hog-nosed snake feeding upon a toad. *Photograph by J. D. Willson*

attempt to swallow them by blowing themselves up like a balloon. Gatersnakes in the western United States have developed resistance to the potent toxins of newts, and are among the only animals that can feed on them. They are among the most colorful gartersnakes, with bright red colors indicating to predators that they have actually accumulated the newt toxins within their own tissues. Mudsnakes specialize on the abundant eel-like salamanders (sirens and amphiumas) found in the dark, muddy bottoms of cypress swamps. Ringneck snakes feed chiefly on the woodland and streamside salamanders that occur in abundance in the eastern United States. Farther west, salamanders are not as common, and there they are more likely to eat lizards.

Most lizard specialists occur in the warm deserts of the Southwest. Lizards are an abundant potential prey in our deserts, and they are the dominant prey of about 23 species of snakes in the United States. Gray-banded and California mountain kingsnakes feed mostly by probing among cracks at night, searching for sleeping lizards. The gray-banded kingsnake pins lizards against narrow cracks to immobilize them for swallowing. Lyresnakes and nightsnakes also prowl the night looking for sleeping lizards, and these have enlarged fangs and venom that help them rapidly kill their prey. Nightsnakes have also been observed feeding on diurnal lizards by ambushing them as they duck under cover. Glossy snakes and long-nosed snakes similarly forage at night, probing underground holes and burrows for sleeping lizards, which they kill by pinning and constriction. Lizards constitute a large portion of the diet of coachwhips and whipsnakes, which actively hunt for lizards during the day and chase them down, and can even catch the remarkably fast whiptails. Some of the unique mountain rattlesnakes of the Southwest feed mostly on lizards as well; these are killed by ambush during the day by the small and attractive twin-spotted rattlesnake, ridge-nosed rattlesnake, and rock rattlesnake. Patch-nosed snakes feed mostly on lizards, although observations of their tactics are rare. Their namesake enlarged snout scale has long been thought to enable them to dig lizards and their eggs from underground chambers. Desert lizards are therefore never safe; whether in cracks, crevices, burrows or in the open—day or night—there are snakes persistently and systematically hunting for them, or lying in wait under cover and hidden among the rocks.

Snakes are excellent predators on other snakes. Our coralsnakes are specialist predators on other snakes, and use their potent venom to immobilize their prey. The previously mentioned short-tailed kingsnake feeds on a single, abundant ground-dwelling crowned snake. Kingsnakes are most famous for their ability to feed on venomous snakes with impunity, and are indeed immune to the venom of copperheads, cottonmouths, and rattlesnakes, which they seize and constrict before eating. Snakes make up the largest proportion of the diet of the eastern indigo snake, which is able to feed on venomous species as well. We will examine these snakes in more detail as predators of snakes in chapter 7.

Several snakes that feed on lizards and snakes also feed on their eggs, and three kinds of snakes are specialists that feed mostly on the eggs of turtles, snakes, and lizards. The scarletsnake of the sandy Gulf and Atlantic Coasts burrows underground in search of reptile nests and has been found in the nest chambers of 10 kinds of reptiles. They feed by chewing open the eggs and greedily slurping out the contents, sometimes with so much enthusiasm that they bury their face in the eggs. Two species of leaf-nosed snakes of the Southwest likewise feed on the eggs of snakes and lizards, and use their enlarged rear teeth to slash open the eggs so they can suck out the nutritious contents. Several other snakes feed on the eggs of reptiles and birds, and eggs are a small but important component of the diet of many snakes. Common kingsnakes relish turtle eggs when they can find them, and colubrid constrictors like king-, rat-, corn-, fox-, gopher-, and pinesnakes never pass up the opportunity to raid the nests of birds, including chickens.

While no American snake is a bird specialist, the feeding activities of our tree-climbing snakes deserve special mention. These and other snakes are also a considerable hazard for ground-nesting birds. Birds, their nestlings, and their eggs constitute an important secondary prey source for king-, rat-, corn-, fox-, gopher-, whip-, and even rattlesnakes. They would probably make up a larger percentage of their diet, but birds are simply hard to catch. But ratsnakes are good at surprising birds in trees, and even better at seeking out and raiding nests. Birds have developed many tactics to counteract these snakes. Most remarkable are the habits of the endangered red-cockaded woodpecker of the southern pine forests, which is the only American woodpecker that excavates its nest cavity in live pine trees. They select old trees succumbing to heart rot, which are softer but still bleed copious strings of sap. The woodpeckers drill sap wells along the perimeter of the nest cavity, which results in a slick, protective patch. Nest-raiding ratsnakes and cornsnakes—both of which are common in the pine savannahs preferred by the woodpecker—can climb up the completely straight boles of pines, but lose their grip at the nest entrance and fall to the forest floor. Some are still able to raid the woodpecker nests, but it is thought that this and many other bird-nest specializations are in large part responses to the acrobatic climbing abilities of snakes.

Snake lovers often make the bold and understandably optimistic claim that most snakes eat rodents. This is a shrewd, apocryphal public relations strategy, and while it is true that some of our snakes are small-mammal specialists, they are actually in the minority. As we have seen, most American snakes feed on invertebrates and ectothermic prey such as amphibians, lizards, and other snakes. Less than 20% of American snakes are small-mammal specialists. These include the boas and most colubrid constrictors, which primarily use the raiding strategy to capture rodents and their babies in burrows. There they are quickly constricted and swallowed. Entire families of rodents are often killed this way when snakes discover a female mouse and her litter. The

A gophersnake feeding upon the eggs of a curlew. *Photograph by Hugh Kingery*

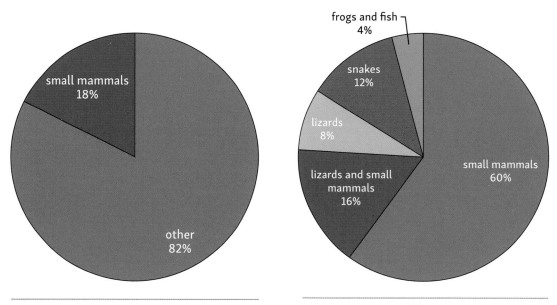

Percentage of US snakes that prefer feeding upon small mammals.

Dietary preferences of the Viperidae.

same climbing snakes that harry birds are also able to scale cave walls to feed on bats.

While in general most American snakes don't eat rodents, it is true that a majority of our venomous snakes do. These snakes use the mobile ambush strategy: seeking out good ambush positions based on their ability to smell rodent scent trails, and then lying in wait for hours, hoping to surprise small mammals with their potent bite.

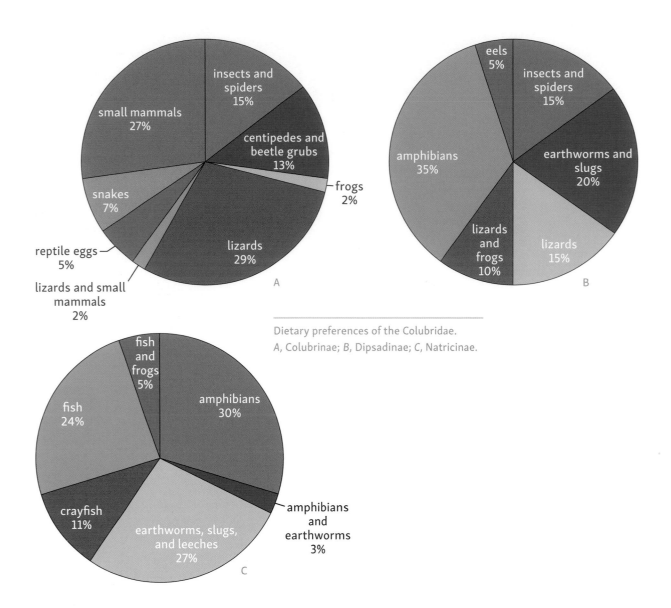

Dietary preferences of the Colubridae.
A, Colubrinae; *B*, Dipsadinae; *C*, Natricinae.

There are several variants of this effective strategy, and certain individual snakes seem to favor one over the others, while others might use all three at various times. Possibly the most common is using habitat edges as ambush sites; rodents use logs as runways along the forest floor, so rattlers are often observed waiting next to logs. Cottonmouths waiting along the water's edge are another example. Timber rattlesnakes and perhaps many others also lie in a similar S-shaped posture among the leaves of the forest floor. With the log-oriented strategy they are more likely to catch deer mice and chipmunks, which run along logs. Among the leaves they are more likely to catch shrews and mice that do not use logs as runways. Rattlesnakes also occasionally prop themselves up at the base of trees, which is an effective strategy for catching tree squirrels and birds. There is fascinating new evidence that rattlesnakes will prepare their ambush site for maximum effect; they purposefully reach

Pitviper ambush positions. *A*, timber rattlesnake oriented toward a log *(photograph by Michelle Byrne)*; *B*, timber rattlesnake oriented toward a tree trunk *(photograph by Dirk Stevenson)*.

out and smooth the vegetation to clear a firing lane. Some rattlesnakes, especially sidewinders and prairie rattlesnakes, lie in wait inside the entrance of rodent burrows, with their heads facing out. This must be a terrifying experience for the rodent: imagine returning home from work to find a tiger waiting behind your front door. Although this may seem a ridiculous scenario, at least one person has fallen prey to a tiger this way.

Baby copperheads, cottonmouths, and some rattlesnakes employ an additional remarkable strategy. They are born with brightly colored tails—usually sulfur yellow or greenish yellow. Although their body color is often much brighter than in adults, young pitvipers still blend in quite well with their surroundings. When they wave their tails slowly and rhythmically, it looks like a brightly colored caterpillar wriggling upon the leaves of a forest floor, which can attract small prey like frogs and lizards within easy striking distance.

While the strategies employed by pitvipers are impressive, so are the contrivances by which rodents avoid becoming the meals of snakes. These countermeasures represent a classic example of what is known as an evolutionary "arms race"—for each improvement made by a predator, prey species develop slightly better defenses, and for each of these, predators must stay one step ahead. Only those predators that avoid being one-upped avoid starvation, leading to ever more formidable predators over the generations. Only those prey that avoid being killed leave progeny, leading to ever warier prey over the generations. Similar evolutionary strategies can be seen in the speed of cheetahs versus the vigilance of impalas. And the sculpted wings and flocking of doves against the fighter-jet speed of the peregrine falcon.

Despite the potency of rattlesnake venoms, ground squirrels in California survive strikes because some populations develop resistance. In areas with rattlesnakes, the blood of ground squirrels is capable of inactivating venom, and they take much longer to succumb to bites. In regions without rattlesnakes, ground squirrels are highly susceptible to the same venom.

Incredibly, rodents also use visual, scent, thermal, and sound signals—all the senses favored by snakes—in an attempt to foil their serpent enemies. Colonial squirrels are constantly wary of the dangers of rattlers and alert their neighbors when they discover one hiding in ambush. They raise alarm calls and jerk their tails in an alarm display (called "flagging"), run in and harass the snakes, and kick sand in their eyes. Nearby squirrels join in, making sure they locate the snake themselves. Snakes discovered in this way quickly retreat from their ambush site and begin looking for a new one. Thermal imagery of ground squirrels engaging rattlesnakes revealed another remarkable defensive tactic: when flagging at rattlesnakes, blood flows into the squirrel's tail, creating a warm glow. This is a signal only rattlesnakes with their infrared pits can see. It probably serves as a decoy to draw the rattlesnake's strike toward their furry, bony, invulnerable tail. When squirrels find old, discarded snake skins, they chew on them and rub themselves all over. Rattlesnakes presented with squirrels using this "scent mask" seem not to recognize them as prey. Kangaroo rats use their big feet to communicate with each other, rapidly thumping them on the ground to produce subsonic sounds. Kangaroo rats use this "drumming" behavior to defend their tiny territories throughout the western United States. They have gigantic auditory chambers that enable them to hear the sounds, which are otherwise undetectable to humans. But snakes hear subsonic sounds that travel through the ground quite well, so it follows that snakes might possibly hear the drumming too. Rulon Clark discovered that Kangaroo rats who find rattlers in their territories also direct their drumming toward the snakes, rapidly circling them, thumping, and kicking sand at them. Snakes revealed in this way emerge from their hiding places and leave the area.

The Role of Snakes as Predators

Predators have important effects on prey populations, but usually herbivore populations are self-limiting, or limited by their own food supplies. Top predators such as wolves and mountain lions have disproportionately large effects on the ecosystems they sit atop, but for smaller predators the effects are less obvious. In a few clear-cut cases, exotic species colonize regions and inflict heavy predation pressure upon naive prey. Brown treesnakes in Guam and Burmese pythons in Florida provide clear evidence that snakes can dramatically affect prey populations. These invasive snakes will be discussed in chapter 10.

Individually, native American snakes don't eat as many prey species as similarly sized mammals and birds. Their low metabolism allows them to go long periods without food, and their efficient conversion of food to body mass enables snakes to acquire the energy needed for growth, maintenance, and reproduction from a fraction of the food required by endothermic predators. A rattlesnake may eat only twice its body weight in rodents per year to fulfill all its needs. This might constitute a mere 12 to 15 kangaroo rats or deer mice per year per snake. A gartersnake may require only 20 frogs a year. This can be compared to a weasel or similar-sized mammal, which might eat that many rodents and frogs in two days.

But snakes often have higher densities than mammalian and avian predators, so as a population, the effects of snakes as predators may be considerable. For example, if weasels are 100 times more effective as predators on rodents than rattlers, but occur at low population densities, then snakes that are 100 times more densely packed would have the same effect on prey. This appears to be the case in many habitats, but in some places snakes may be so uncommon that their role in the ecosystem may be admittedly weak.

Snakes may have big effects on rodent populations when warm-blooded predators can't do the job. Many rodents undergo predictable population fluctuations. These are often tied to decade peaks of rainfall, and mammalian predators often respond to bumper crops of prey by feeding and reproducing

Western diamondback rattlesnake feeding upon a desert cottontail. *Photograph by Brian Sullivan*

Eastern diamondback rattlesnake feeding upon an eastern cottontail. *Photograph by Dirk Stevenson*

more. These great years are often followed by prey population crashes, however, which in turn cause predator populations to crash. These often occur regularly once a decade or sooner, and are called "boom-bust" cycles. Given their exceptional abilities to suffer through lean years, snakes likely constitute a significant predation pressure on the recovering rodent populations during the bust, possibly forestalling their next boom.

While all snakes are predators, the extent to which they fall prey to other species is also an important ecological consideration, and is the topic to which we turn next.

7 · Snake Eaters

To be born a snake is to be thrust into a place a-swarm with formidable foes.

STEPHEN CRANE, *The Snake*

A Brutal Murder in West Texas

The west Texas grassland is tawny yellow, stretching gradually until it disappears into desert to the south at a point too far to see and into the mountains a day's walk to the north. The sky is thin, hazy blue with a few insignificant puffs of cloud and one thunderhead the size of a city to the east.

A snake makes its way back to its burrow after a night's hunt but tarries because the morning sun is so nice. Without warning, a small bird lands hard on the snake. The bird is smartly dressed like a contract killer in gray, black, and white. The bird savagely pecks the snake as it writhes on the ground, then it flies off with it over to a barbed-wire fence. The bird uses its cruel, hooked beak to hammer the snake down onto a singular barb. It pounds the snake's neck down over the barb 10 times, then 20, the snake writhing in agony. The snake now firmly crucified, the bird grips its head with the hooked beak and rips it off.

Being a farm girl from Australia, my wife can't go long without needing red meat and starts getting light headed and antsy if she hasn't had a steak in a few weeks. So when I told her about the job in west Texas, I promised her we'd be living in beef country and I knew just the place to take her when we went down there for my interview.

A western groundsnake crucified by a loggerhead shrike in west Texas.
Photograph by Crystal Kelehear

La Kiva had become moderately famous because it was featured in *GQ Magazine* as "The #1 most bizarre bar you must visit before you die." I knew it from previous visits to the Big Bend country. Driving out of the west entrance of Big Bend National Park, the road descends into barren badlands wasted further by quicksilver mines. You come to Study Butte at a highway intersection with a gas station, a diner, and a few other eclectic businesses—small grocery, rock shop, and chapel—set deep in a mess of canyons and enormous brick-colored bare mountains shaped like cabbage. Head west farther along Route 117 and you cross Terlingua Creek, a pale draw of dry white sand and chalky gypsum clay and a small ellipse of stagnant water. On the west bank is La Kiva—a small adobe flanked by planted palms strung with Christmas lights with a big, barren gravel parking lot.

You used to go through a heavy wooden castle door and descend stairs into a cool, dark dungeon corridor lined with medieval torches. The outside always being saturated by the great Chihuahuan Desert sun, it took a while for your eyes to adjust. When they did, you beheld a handsome wooden bar set within a socket of rock. To the right of the bar was a reconstruction of the ancient creature *Penisaurus erectus*, a strange beast with horns and hooves obviously made of old cow bones. To the left of this fossil and behind the bar invariably stood Glenn Felts, the owner of this fine establishment.

Glenn had a boyish face but shoulder-length dirty brown-gray hair. He was missing a few of his front teeth, but that never prevented him from smiling. He typically wore a black heavy-metal T-shirt and a trucker hat, so he looked more like a roadie for Guns n' Roses than he did a Texas barkeep. He was always friendly with tourists, who sat at tables, but talked mostly with the locals, who sidled up to the bar. It was rumored that he was the local connection if you needed drugs. These things never interested or concerned us, because we were there for the cheap lager on tap and the only thing on the menu: a 9-ounce sirloin served with a side of pinto beans, coleslaw, and a rolled tortilla.

After a day hiking in the Chisos Mountains, we arrived in Terlingua anticipating this juicy steak. Instead, to our great disappointment, the gravel parking lot of La Kiva was roped off. We drove up the road a little farther to the first restaurant we could find, a place called High Sierra, a two-story wooden building aiming to look like an Old West saloon. We were soon draining tall glasses of Shiner Bock and waiting on bison burgers. The 2,000-foot cliffs of Santa Elena dimmed from tan to pink to the southwest, and the Chisos Mountains were darkening in their entirety to the east. While we waited, the night's musical talent began setting up: a white-haired old hippie, with the stage name Dr. Fun. I shot the breeze with Dr. Fun for a few minutes, asking him if he played covers and then responding that I was glad he didn't. Then I asked him why La Kiva was closed. He closed his eyes, tilted his head down, and shook it.

"Oh man, it's terrible. The owner . . . he was brutally murdered," he told us with a Texas accent that wasn't quite thick enough to belong out here but that he could easily have picked up in Houston. The word "brutally" he said with some emphasis and perhaps even theatrics, the syllables bumping off slowly like a rock plunking down a hill, the first consonant hanging for effect. "They said that based on his injuries the guy must have been beating on him for over an hour after he was already dead."

He volunteered a few more details about the incident while I listened with morbid fascination. It had only happened a month or so before, and the suspect—a local river guide beloved by the community—was in custody. The murder was presumed to have been drug related and aided greatly by alcohol. Glenn's body was discovered in the morning by an employee who came in to work the day after the murder. She noticed somebody lying face down in the gravel parking lot, but this did not concern her. To quote Sheriff Ronny Dodson, "It's a bar," and seeing a person passed out in the parking lot was nothing unusual. Instead, that the lights were still on and the house music was still playing inside were the only indications something was wrong. She went all the way over to Glenn's house to look for him and, nothing doing, finally realized that the guy in the parking lot was Glenn.

There is just something about west Texas that gets people killed in unusual and dramatic ways. The preface to the book *Death in Big Bend* assures the reader that in Big Bend National Park you are most likely to die of a car accident or heart attack—just like anywhere else. And that is true. But what follows is a series of crackling tales about unsolved murders, rock-climbing expeditions gone amok, and stupid tourists dying of thirst in horrible and preventable ways. It's as if the desert doesn't really want you, but when it decides to take you, it does it with some flair. Maybe this is why Cormac McCarthy set so many novels in these borderlands. The movie adaptation of his book *No Country for Old Men* was filmed in west Texas. A character says at the cryptic ending of the movie: "What you got ain't nothin' new. This country's hard on people. You can't stop what's coming."

I've always loved it out here. Big Bend National Park has long been my favorite. I first heard about it in a sensationalist article from *Outside Magazine* about "America's Most Dangerous National Parks." The article described illicit border trafficking and a triple murder, and I was hooked. I love the juxtaposition of the mountains, the desert, and the river. From the floor of the shrubby Chihuahuan Desert rise the Chisos Mountains—a craggy range of collapsed volcanoes and lava flows covered with scraggly oaks, pinyons, and junipers. There is a fascinating history of biotic and cultural interchange: the northernmost populations of animals and plants otherwise confined to Mexico are found here. Likewise, until recently, there was a warm and porous border—you used to see Mexicans drive their pickups across the river to fill up drums of fuel because it was cheaper on the Texas side. When you talked to the rangers about crossing the border, they would say they were obliged as federal employees not to condone illegal crossings but that personally they thought the beer over there was pretty cold.

And there are wonderful snakes here. Within a few hours of driving after finishing at High Sierra, we were giddy, having found nightsnakes, ground-snakes, western diamondbacks, and the evening's prize: a Trans-Pecos rat-snake. It was a handsome, five-foot-long blonde snake with a characteristic pattern of black H's down the back. The face was rounded and attractive, but what caught my attention most were the smoky blue eyes: expressive, inquisi-tive, mysterious. They feed on small rodents but are also accomplished bat kill-ers: they can scale sheer rock crevices to surprise resting bats. The Trans-Pecos ratsnake is one of the area's specialties.

Portrait of a Trans-Pecos ratsnake. *Photograph by Curtis Callaway*

I got the job, and my wife and I began weekend ventures out into the surrounding country. After-noon drives revealed new and wondrous crea-tures, each belonging unwittingly to the terrible web of murder and carnivory of west Texas: com-mon blackhawks circling near the Rio Grande; pronghorn males grappling exhaustively during the rut—these contests can be to the death; a desert massasauga with two lizards in its belly; a gray fox in the headlights ducking into a bush across the street; a pair of juvenile bobcats prowl-ing the highway; a roadrunner gliding to the top of a shrub with a lizard in its beak; a family of kit foxes lounging around their den. Mountain lions still pace here, and until recently wolves were still present. Texas's only grizzly bear was shot over 100 years ago in the Davis Mountains.

But snakes are the subject of an inordinate number of conversations around the tiny towns dotting this primitive landscape. The word "rattlesnake" echoes faintly across yards like a crow's call. Introductions and hat tips and handshakes are quickly followed by relayed sightings of rattlers, because the weather is too permanent for much conversation.

One afternoon we drove across the vast Ryan Flat, a rich desert grassland west of Marfa. We slowed and turned down a smaller road. Right away we saw something dangling on the barbed wire. It was a pitiful lizard, sun bleached, headless, and mummified by the desert sun. We parked the car and walked both sides of the road. Along a mile-long stretch of Texas farm road we found 35 grasshoppers, 46 beetles, 23 lizards, 1 small bird only identified by the pres-ence of feathers and a foot, and 3 snakes, 2 of which were sun dried and uniden-tifiable. All of these were killed and left there by a loggerhead shrike, a bird known for impaling its prey on sharp objects.

One was headless, flattened with age, and textured like jerky, but it was clearly a groundsnake—a small, variable-colored species of the Southwest and plains states. In life, it would have been a salmon orange with black bands. When you find these little artifacts—tiny harmless creatures painfully pinned and slowly dismembered by a purposeful and rather professional killer—it is difficult not to imagine what their final minutes must have been like. The heads of horned lizards hammered onto the barb, impaled through their mouths. Lizards pierced through the armpit, with agonizing looks stamped permanently on their faces. Black-banded orange snakes, decapitated but coiled around the barbed wire as if still trying to get away.

West Texas is wonderful. So is nature. But they are both cruel.

Snakes occupy an unenviable position in the food webs of North American forests, prairies, deserts, and swamps. Even though snakes are predatory, snakes themselves are food items for a variety of larger predators. They occupy a teetering location dead in the middle of the food chain. Most people probably don't like the sound of eating a snake, but for many, if not most animals, snakes are the perfect food. They are long and skinny, so they are an easy-to-swallow tube, and they are practically all meat. If you can find them, they are rather slow and easy to catch. In America, most snakes aren't venomous and so pose little danger to a larger animal.

As ectotherms, they have slower growth rates than mammals or birds. Even large species stay smaller longer than would a mammal of similar body size. And a small body size immediately makes any animal more vulnerable to predation. Snakes must regulate their body temperature through behavior, so there is always the chance that a warm-blooded predator might catch a snake when it is cool and vulnerable. The metabolism of a snake does not support prolonged bouts of endurance, so predators can quickly exhaust snakes during a chase. These constraints conspire to make snakes easy prey. They must negotiate a busy intersection in the food web between the small prey items they eat, small predators that can eat small snakes, and larger top predators that feed on all comers.

Practically every predatory animal you can think of has killed an American snake. Black widow spiders, fishing spiders, tarantulas, scorpions, centipedes, fire ants, carpenter ants, giant water bugs, Jerusalem crickets, beetles, crayfish, and crabs are among the arthropods that have been discovered snacking on snakes, although these invertebrates are known mostly to feed on young snakes. We don't know how many snakes are killed by such creepy crawlies because most of this knowledge comes from chance observations. But it is likely a significant source of mortality: baby snakes are hatched of such a puny size (sometimes as small as an inch or two and not much wider than yarn) that these often-vicious invertebrate hunters probably do not consider them much more difficult to dispatch than a worm. Some are far more formidable than a baby snake: the hideous, nearly foot-long centipede *Scolopendra heros* is a known predator of the groundsnake, whose average body size is smaller than the centipede's. The threadsnakes of the desert Southwest frequently fall prey to the fearsome arthropods found there. They are victims of centipedes, black widows, and one was discovered being hauled across a dune in the clutches of a large, hairy scorpion that had stung the poor snake to death. Venomous snakes are not spared from these leggy predators either; a dead cottonmouth was found in the pincers of a ghost crab near Virginia Beach. Perhaps the strangest invertebrate predator known for an American snake is the octopus, which is a known predator of the yellow-bellied seasnake. Watersnakes sometimes even manage to die after getting trapped between the valves of freshwater mussels.

Small snakes also fall prey to fish. Our knowledge of these predation events comes mostly from occasional impromptu dissections rather than systematic studies, so here too the significance of these predators is poorly known. Most records also come from large game fish, which are more likely to be caught and dissected. Most predatory fish simply gulp their prey down, so small snakes caught swimming are swallowed whole. Large predatory fish such as large-mouth bass, gar, and trout are known predators of snakes, especially semi-aquatic species like watersnakes, gartersnakes, and even cottonmouths. But terrestrial species also fall victim to fish: brook trout have eaten sharp-tailed snakes, and brown trout have eaten ringneck snakes and gophersnakes. A brown trout was even caught with a prairie rattlesnake in its gullet. The yellow-bellied seasnake falls prey to diverse marine fish such as puffers and snappers.

Large frogs have no problem swallowing smaller individuals of even the most dangerous snakes. The most prolific snake killer of the amphibians is the American bullfrog, which has consumed just under 20 species of American snakes, including the copperhead, pygmy rattlesnake, harlequin coralsnake, and both the eastern and western diamondback rattlesnake. Its appetite for snakes as well as other small animals makes the bullfrog a threat to species in places where it doesn't belong: its natural range is in the eastern United States as far west as Texas, but it was artificially introduced in some of our western states. There it is an invasive species and has unfortunately fed upon some of the rarer gartersnakes in California, such as the giant and two-striped gartersnake.

Rounding out the list of known amphibian predators of snakes are some of our giant salamanders—strange, slimy beasts unfamiliar even to some biologists. Witness the hellbender, a great 3-foot salamander of cold Appalachian streams that is as ugly as it is wonderful. Although crayfish make up the bulk of the hellbender's diet, a queensnake fell prey to one. The giant eel-like salamanders of the southern swamps are known to feed on snakes too—a greater siren has eaten a brown watersnake and striped swampsnake, and the two-toed amphiuma is a known predator of the glossy swampsnake.

Lizards should probably be considered a minor threat to most snakes, as they are much more likely to be eaten by snakes than to be found feeding on them. An exception is the collared lizard, a large, fleet killing machine that sometimes runs so fast it lifts itself off the ground to run on two legs. Collared lizards have unusually large heads that allow them to feed on smaller lizards, but they are also known to feed on ringneck snakes, groundsnakes, gartersnakes, rough greensnakes, and one even ate a rock rattlesnake. Turtles get in on the action; the common snapping turtle has fed on three different kinds of watersnakes, a ratsnake, and the cottonmouth. The Sonoran mud turtle of the desert Southwest seems to have a peculiar knack for killing serpents; it is a poorly studied species, yet two kinds of snakes are known from its diet.

With their giant size, thick leathery skin studded with bony plates, and crushing jaws, alligators are efficient and nonchalant predators of snakes. They kill large adult snakes of several semiaquatic southern species, including the cottonmouth. Alligators are probably one of the few predators that can handle large adult cottonmouths and eastern diamondback rattlesnakes with no problem at all. They do not simply rely on their thick skins, either. Alligators snap down on snakes and kill them with shaking movements so vigorous that the snakes are helplessly tossed from side to side. This probably breaks their spinal cord in several places. They then gulp them down in one piece.

Ironically, snakes are the reptiles responsible for killing the most snakes, and they are the only animals in the United States that specialize on eating snakes. The snake eater with the most particular palate is the short-tailed kingsnake, a unique, exceedingly thin, orange-and-black blotched snake of the sandy midsection of Florida. This snake's diet is made up of only one type of snake: the Florida crowned snake. These they seize either on the head or rear and

Florida kingsnake eating a gartersnake.
Photograph by Ian Deery

begin swallowing head-first. Larger individuals are clumsily constricted for some time before being swallowed alive. Other snakes with a penchant for serpents include our three coralsnakes—the harlequin, Texas, and Sonoran coralsnakes. These cobra relatives are known to feed chiefly on snakes, although they also occasionally eat small lizards. The harlequin and Texas coralsnakes mostly feed on small, helpless snakes such as crowned snakes, earthsnakes, gartersnakes, and brownsnakes, but they have also taken small individuals of more formidable species like rattlesnakes and copperheads. Similarly, the Sonoran coralsnake of the Arizona desert feeds mostly on small burrowing snakes like threadsnakes, groundsnakes, and leaf-nosed snakes. The coralsnakes have potent venom useful for quickly immobilizing their prey; most of the small snakes they eat are probably rapidly paralyzed by their chewing bite and don't stand a chance.

Many other snakes are far less choosy about what they eat and have more generalized palates, but they include many snakes in their diet. Examples include the coachwhip—the alert, athletic, and swift snake of the southern states—whose diet ranges from grasshoppers to birds and includes many kinds of nonvenomous and venomous snakes. As we learned in chapter 6, the cottonmouth—the stout haunt of southern swamps—will eat nearly anything, and snakes are certainly not spared. During my studies of this expert hunter, I found two cottonmouths in the process of dispatching one of their fellow serpents: one I found with the last half of a brown watersnake hanging from its mouth, and the other's massively distended belly contained a close relative: a partly digested copperhead. Cottonmouths ambush other snakes as they crawl along the water's edge, snatch them in their jaws, and inject them with their disintegrating venom. The struggle between the cottonmouth and copperhead was probably a savage contest, but clearly the larger cottonmouth prevailed.

The kingsnake is perhaps the most famous snake killer, although this species' diet is not as narrow as you might have thought. They also feed on a variety of other vertebrates, such as small mammals, lizards, frogs, turtle eggs, and many others. But they are enthusiastic killers of snakes, and seem to have a real preference for certain venomous species. They are also the best equipped for killing venomous species: their blood contains a protein that somehow renders venom inert, so they can grab and constrict large venomous snakes with impunity. They are the most accomplished snake killers among American snakes, having fed upon 40 different species.

The effect of these snakes on populations of other snake species is now becoming clear. Eastern kingsnakes have recently undergone a dramatic decline in some southern states, and they are now absent from several areas where they were formerly common. Instead, copperheads—once considered fairly uncommon in the same areas—are now the most common snakes. A thorough comparison of the abundances of both species throughout the South-

Eastern kingsnake eating a cottonmouth. *Photograph by Pierson Hill*

California kingsnake eating a Panamint rattlesnake. *Photograph by Drew Kaiser, courtesy of Death Valley National Park*

east found that more kingsnakes typically means fewer copperheads. And in the places that now have few kingsnakes, copperheads are on the rise.

Indigo snakes are probably the most skilled and ruthless snake killers in America. The eastern and Texas indigos are the largest North American snakes and have no trouble feeding on smaller snakes by swallowing them alive. But they also feed on large venomous species—for the eastern indigo, a 3- or 4-foot eastern diamondback is not out of the question. Likewise, for the

Texas indigo, a large western diamondback is of no consequence. But unlike the kingsnake, it is unknown whether the indigo snake is immune to the venoms of these pitvipers. What is clear is that for the indigo it doesn't matter: they are agile and powerful snakes that simply seize these dangerous species by the face and savagely chew the head with their robust jaws. Once the head is pulverized, they can then swallow the dead rattlesnake without remorse.

Mammals are the most numerous potential predators of snakes, and North America boasts a variety of vicious carnivorous mammals that could potentially dispatch snakes. But there are also a number of small, but no less vicious, mammalian snake killers. Principal among these are the shrews and moles. Their high metabolism rivals that of a hummingbird, and they must feed constantly or face rapid starvation. Shrews end up eating three or four times their weight in soft-bodied invertebrates each day. Larger prey items are also taken; certain species, such as the short-tailed shrew, are adept hunters of frogs, salamanders, mice, and snakes. These they kill with blinding attacks, parries, and jabs with their tiny, hideous, peg-like teeth. They can kill snakes much larger than themselves.

Watching a shrew kill a snake resembles the athleticism of the fierce Tasmanian Devil from the cartoons. Their movements are whirling, rapid, and jerky; in one moment they are gnawing on the tail; in the next they pierce the neck; in another moment they do a backflip; seize the midsection, scurry to the head, and bite the neck again. They deliver the proverbial death by a thousand cuts, and for their prey it must be a terrifying and horrible way to go. The short-tailed shrew even has an additional adaptation that should ring familiar to snakes: they're among the few venomous mammals known, conducting toxic saliva into their prey. Shrews and moles are known to feed on at least 11 species of snakes but probably consume many more species, especially small-litter snakes and small individuals of larger species. Many of these will turn the tables on shrews when they grow up; shrews are common prey items for most snakes that feed on small mammals.

Rodents also kill snakes on occasion. Even large snakes that favor rodents as prey can be killed; rock squirrels and California ground squirrels are particularly adept at harrying gophersnakes. They dash forth like a mongoose and relentlessly gnash at the snake while it tries to crawl away. Adult gophersnakes are sometimes found dead with several chunks removed from their backs after encounters with rodents. In 1921, a college professor in Texas witnessed a brown rat take on a small western diamondback rattlesnake. After a 10-minute battle, during which the snake struck repeatedly and missed the rat every time, the rat "won the battle by a single gash of its sharp incisors into its head, the snake wilting instantly."

With the widespread availability of digital cameras and camera phones, rodent attacks are becoming easier to document; several YouTube videos depict rock squirrels attacking snakes. The observers are usually flabbergasted that the squirrels are winning the battle. In one, a bystander is surprisingly sympathetic to the snake, which appears to suffer greatly. In many cases the aggressive attacks are not intended to be predatory but are simply defensive measures on the part of the rodent. But rodents that get the upper hand on snakes will press this advantage to its fullest, and the result for the snake is the same whether the rodent intends to eat the snake or not. Small snakes similarly fall prey to rodents like chipmunks, squirrels, and mice, which are often far less vegetarian than you'd expect; they can be quick and effective killers of insects and small vertebrates. When compared to the sheer number of rodents consumed by adult snakes, however, the few cases where the roles are occasionally reversed are relatively insignificant.

Similarly, large ungulates kill many snakes out of what appears to be a sense of self-preservation or some ancestral hatred. Deer, pronghorn, domestic goats, horses, and cows are all known to trample snakes to death on purpose. Their rock-hard hooves, stiff, bony limbs, combined with thick hides, are ample protection from snakebite. When they spot the snake, hoofed animals prance upon snakes multiple times, usually by holding their two front legs stiff and plunging them downward while jumping, resulting in a badly mutilated snake. This was described by many eyewitnesses in Klauber's exceptionally comprehensive two-volume set about rattlesnakes, but the best was related by Albert Madarieta, of Oakley, Idaho: "I have seen buck deer kill rattlesnakes on three different occasions. They just back up and run and stomp them: then they turn and do it again. After a buck deer gets through, there ain't a piece of rattlesnake over one inch in length. A buck deer really makes sure that they're dead."

America's only marsupial mammal is known to feed on snakes, and may have a special adaptation that allows it to feed on venomous species. The opossum is a decided opportunist when it comes to food, feeding on any edible item available, from persimmons to chicken eggs. They are known to have consumed at least 12 different kinds of snakes, which they probably find using their excellent noses. Opossums amble along the forests and meadows of the eastern states (they are now also introduced along the Pacific Coast), sniffing any likely pocket or hole for possible prey. They have a battery of sharp, peg-like teeth—indeed, more pairs of teeth than any terrestrial mammal—with which they can easily dispatch most small vertebrates. But their best weapon for dealing with snakes is their blood. The blood of opossums is a medical curiosity; they are somehow practically immune to the horrifying rabies virus, and their blood contains proteins that can rapidly incapacitate snake venoms.

America is populated by a diverse array of carnivorous mammals that range in size from the 57-gram least weasel to the 950-kilogram grizzly bear. Probably every species of predatory mammal in the country has fed upon a snake at one point or another. Our understanding of the interactions between these mammals and snakes is slightly better than for other kinds of predators. Thorough studies of mammal diets have been conducted from careful analysis of

Bobcat eating a banded watersnake.
Photograph by Joe Stevenot

scats and gut contents. Gut contents are often examined coincident with predator eradication programs or incidental to harvesting mammals for the fur trade. These studies all agree that snakes are a minor yet pervasive component of the diet of American mammalian carnivores. Snakes throughout the country are documented as prey for such diverse predators as ringtails, raccoons, otters, foxes, bobcats, coyotes, badgers, wolves, and bears. Most of these predators are also known to kill venomous snakes. How exactly they kill formidable venomous species and avoid being bitten is not always known; because predatory mammals hunt at night it, is difficult to observe their behavior. And it is known that sometimes even predators the size of black bears can die from snakebite. For example, a four-month-old black bear in western Virginia was found lying dead on the forest floor. A thorough autopsy revealed swelling and discoloration in the cub's hind limb. A pair of widely spaced puncture wounds indicated that a timber rattlesnake was the likely culprit.

Predatory mammals like foxes and bobcats patrol large areas looking for prey. They are mostly hunting for small mammals like rodents and rabbits, probing burrows and other likely hiding places and occasionally giving chase.

While on their daily rounds, they often surprise smaller animals like insects and worms, which they quickly pounce upon and kill with their sharp teeth. Small snakes belong to this prey category—small food items found and dispatched when opportunity knocks. Larger snakes are also consumed but require more time and perhaps delicate handling in case they are venomous. Perhaps for this reason, snakes are not a staple of the diet of any American predatory mammal; they do turn up as food, but only occasionally.

Snakes make up about 1% to 5% of the prey seen in the scats or stomachs of bobcats studied all over America. But in 23 of 35 separate studies on bobcat diet, no snakes were found at all. Most studies show that small fish and crayfish make up the bulk of the diet of river otters. But two kinds of watersnakes make up just 6% of the diet of river otters in the Atchafalaya Basin of Louisiana, a region teaming with snakes. A meticulous and grave study in 12 western states examined over 3,000 coyote stomachs, of which only about 1,400 contained food items. Of these, only 3% contained the remains of snakes, including six western diamondback rattlesnakes, five bullsnakes, a gartersnake, and a racer. Just 4% of over 200 scats of fishers—a large weasel of the dark coniferous forests of the Sierras, Rockies, and northern states—contained the remains of snakes. The ringtail—a small and unbearably cute relative of the raccoon found in rocky canyons of the Southwest—is known to eat several species of snakes. A remarkable study examined mummified ringtail scats estimated to be 1,700 years old from a cave in the Grand Canyon. These fossil scats included remains of eight of the ten snakes found in the area. But snakes appear to be only a minor component of this mammal's diet as well.

Bobcat eating a gophersnake. *Photograph by Kay Rhodes*

Bobcat attacking and killing a western diamondback rattlesnake at the Rolling Plains Quail Research Ranch, Texas. *Photographs by Susan Cooper*

Black bears eschew snakes. Bears are otherwise well equipped to feed on snakes with their excellent near-focus eyesight, sense of smell, and foraging behavior that includes turning over logs and rocks. Their foraging sign can be mistaken for the work of herpetologists. Yet a snake has never been recorded in any of a dozen analyses of thousands of black bear scats across several US states. Overall, diet studies from around the country suggest that snakes are not considered fine cuisine by carnivorous mammals.

Still, even small numbers of snakes taken by predatory mammals add up when you consider how many of these mammals there are and how frequently they feed. Raccoons are common and densely populated across the country, and they are known to have fed upon 18 kinds of snakes, including formidable species like cottonmouths and timber rattlesnakes. Snakes make up only a minor component of the diet of raccoons, which feed on far more crayfish, snails, frogs, and plant material than snakes. However, even at a small percentage, the impact of these common predators on snake populations may be substantial. Add to this the occasional snacking of coyotes, ringtails, martens, fishers, weasels, bobcats, four kinds of foxes, and badgers, and you get the impression snakes must have to live constantly on their guard.

Many of these predators are also skilled at finding and digging up reptile nests, which may represent a much more significant source of predation for snakes. Raccoons and skunks are especially adept at finding nests, and skunks in particular seem to have a penchant for eating snake eggs. In one of the few detailed examinations of snake nests ever conducted, a large number were destroyed by skunks.

But mammals aren't the primary enemy of American snakes. In terms of sheer numbers of snakes killed each year, birds probably outnumber even people as snake killers. Our understanding of birds as predators of snakes comes from occasional casual observations noted in scientific periodicals, and also more detailed observational studies of the feeding behavior of predatory birds. But our best evidence comes from a few choice studies involving birds returning to their nests with food for their young, where a scientist simply sits in a blind all day watching birds and counting the prey in their talons. These studies reveal the great extent to which birds dispatch snakes. But even birds better known for eating small prey like earthworms and acorns get in on the action. And some rather small birds are remarkably skilled at killing snakes.

Perching birds feed mostly on small seeds, fruit, and insects, but larger species like jays, thrashers, catbirds, mockingbirds, crows, magpies, and ravens are more predatory and also feed on snakes. All told, such passerines are known to have consumed over 20 kinds of snakes—mostly small species like wormsnakes, blindsnakes, gartersnakes, brownsnakes, red-bellied snakes, and ringneck snakes. It is possible that these birds recognize these species as harmless, or, in the case of an American robin that ate a wormsnake, it is possible they mistake snakes for some kind of worm. However, even songbirds sometimes eat rattlesnakes.

One passerine bird deserves special mention. The loggerhead shrike is a pure killer, but it is more closely related to the mockingbird than it is to falcons, hawks, or other birds of prey. Its feet reveal this kinship; they have weak, small claws equipped for perching, not grasping. But the beak of the shrike is strong and hooked, and it has a malicious notch, or "tomial tooth," which it uses to grasp prey firmly. The beak looks like that of a kestrel, and it is used in similar fashion for seizing prey. Owing to the mismatched tools possessed by the loggerhead shrike—a harmless foot and killer beak—the bird learned to improvise. Because it has no talons, the loggerhead shrike instead impales its hapless victims on the first available spine, thorn, brier, or barbed-wire fence. The bird firmly grasps the prey with the beak and nails it down over the sharp spine, often up through the skull, and often while the prey is still fussing. They can then rip off strips of flesh—naturally, while the snake still wriggles cruelly on the barb. Loggerhead shrikes have impaled at least 20 kinds of snakes, including coralsnakes and rattlesnakes. In one fascinating case, a pair of shrikes hunted cooperatively to take on a large snake; they took turns attacking an adult western terrestrial gartersnake before simultaneously attacking the snake. They eventually killed and flew off with it.

The roadrunner is an ambitious killer of insects and small vertebrates, and deserves far more respect than the aloof ditz depicted in the Looney Tunes series. This large ground-dwelling bird is a member of the cuckoo family, a large group of birds famous for laying their eggs in the nests of other birds. But the roadrunner does not take part in such trickery, and instead the mated pair rears their own young. Regardless of their breeding habits, many cuckoos

A loggerhead shrike with its prey: a Dekay's brownsnake. *Photograph by Christopher Cunningham, courtesy of twoshutterbirds.com*

Roadrunner attacking and consuming a threadsnake.
Photographs by Lisa Abernathy

have rapacious killing abilities. Roadrunners specialize in chasing small prey out in the open desert. They have fascinating behaviors for handling dangerous prey, but their method of dispatching smaller snakes and lizards is the picture of simple brutality: they grab prey with their stout beaks and repeatedly smash them to death on rocks.

Larger snakes, including rattlesnakes, are killed with slightly more finesse. The roadrunner spreads its wings wide as a decoy, prancing slowly from side to side to draw the attention of the snake. During such encounters, the snake raises its coils into a standard defensive posture and rattles furiously, trying to keep its face aimed at its tormentor. The roadrunner's wings appear to work simultaneously as a matador's cape and impervious shield. Most of a bird's wing is made up of feathers or stout bone, both of which are materials unaffected by venom. By taunting the snake with its broad wings, the roadrunner can deflect the strike to where the venom's power falls flat. Roadrunners parry and tiptoe in loops around the defending rattlesnake, allowing the snake to strike and miss repeatedly, occasionally folding the wings quickly to run in for an attack. When the snake strikes, the bird bounds several feet in the air. The snake seems to become exhausted by its efforts, and the roadrunner is able to grasp it by the neck and pummel it to death. Roadrunner pairs have been observed tag-teaming rattlesnakes—one distracts the snake with its broad-winged display while the other attacks from behind—and in this way they are able to dispatch larger snakes. Roadrunners seem to know how big a snake they can handle, and simply ignore very large rattlers. At least 14 kinds of snakes have fallen prey to this charismatic bird.

Wading birds like herons, egrets, ibis, cranes, and bitterns feed on snakes, and collectively this group of related birds has fed on about nine types of semiaquatic snakes. They wade along the edges of wetlands spearing almost any animal they see with their long beaks. Then they typically cough down the prey whole with a quick series of jerks that position the prey straight down the throat. Even spiny food like sunfish can slide down the throat of wading birds. Snakes struggle helplessly in their tough beaks, as if caught between blunt scissors. Small venomous snakes are defenseless if they can't reach past the solid and impervious beak, and it is likely such birds do not attack snakes large enough to reach their heads. Cottonmouths are probably killed most frequently by these birds: two kinds of ibis, three kinds of herons, sandhill cranes, and wood storks are known to feed on this snake. However, these birds are surprisingly minor predators of snakes. Snakes make up only a small percentage of the total diet of wading birds. Hundreds of herons of four species nesting on the edge of Lake Okeechobee in Florida were observed for three years during the nesting season. They returned to nests bringing their young thousands of fish, hundreds of aquatic insects, tadpoles, and reptiles like green anoles and five-lined skinks. A single black swampsnake was the only snake brought back to the nests.

Whereas carnivorous mammals and wading birds may feed upon snakes only occasionally, birds of prey are major predators on American snakes. Snakes are killed by ambush from above, the sharp talons pinning the snake before it is killed with a fatal blow to the neck using the hooked beak. Raptors then prudently remove the head and discard it. A red-tailed hawk in Arizona was seen using an ingenious killing method: the glossy snake in its talons dangled below while the bird swooped close to a boulder and at the last moment

Swainson's hawk with a gartersnake.
Photograph by Mike Ross

American kestrel capturing a racer.
Photograph by Ron Dudley

pulled up in flight. In this manner the bird slapped the snake against the rocks several times until it was quite dead. With few exceptions, most kinds of raptors have taken snakes, and for some, snakes represent a sizeable percentage of the diet. This is no more apparent than during the nesting season, when dietary studies demonstrate how frequently snakes are brought back to nests.

Red-tailed hawks are widespread, common raptors found from coast to coast. They are generalist predators that feed on practically anything that moves: small mammals, birds, amphibians, reptiles, and larger insects. No fewer than 35 species of snakes have fallen prey to them. Not only do they feed on numerous kinds of snakes, but there is also evidence that they feed on them regularly. Six red-tailed hawk nests were observed along the Columbia River in Oregon during the nesting season. During just six weeks, an astonishing number of snakes were delivered to the bird's youngsters: 32 yellow-bellied racers, 27 gophersnakes, 1 gartersnake, and 2 northern Pacific rattlesnakes. These represented about half of the prey items captured in terms of weight. Given that red-tailed hawks are the most common predatory bird in North America, and millions nest every year in every US state, you can imagine the kind of damage red-tailed hawks can inflict on snake populations.

Several other raptors feed on snakes with high frequency. Although it is less widespread than the red-tailed hawk, the red-shouldered hawk of the eastern United States feeds on snakes and other ectothermic prey more routinely. Twelve different studies from all over the bird's distribution indicate that reptiles make up about 20% of their diet; of the reptiles eaten, snakes are typically the most important in terms of prey weight. The list of snakes fed upon by this bird is impressive: 16 species, including eastern coralsnakes. Venomous pitvipers are curiously absent from the list of prey species. The swallow-tailed kite is a gorgeous, lily-white raptor of the southern swamplands and surrounding agricultural fields—it flies in graceful arcs, its gray back trimmed in black, with a sleek pair of pointed tail tassels. They are often illustrated flying with a rough greensnake hanging from their beaks, and they do indeed have a knack for spotting and killing these camouflaged tree-dwellers. Swallow-tailed kites nesting along Lake Okeechobee bring back about four snakes per hour to their nests. This is an impressive number; herpetologists would consider it a good day to find four snakes an hour just about anywhere. The southern states are home to additional birds of prey that feed on snakes, apparently with more frequency than more widespread raptors found in the central and northern states, which feed mostly on rodents. These include the white-tailed hawk, gray hawk, common black hawk, and zone-tailed hawk.

During the day, raptors as diverse as golden eagles, northern harriers, American kestrels, and several kinds of hawks patrol the air with their incredible eyes scanning the grasses, leaves, and swamp edges looking for small vertebrates. In the evening, they take up a roost and sleep. Then it's time for the night shift.

Red-shouldered hawk killing a checkered gartersnake. *Photograph by Ann Mallard*

Red-shouldered hawk consuming a striped crayfish snake. *Photograph by Daniel Wakefield*

Several kinds of owls are known to feed on American snakes, although none add as many snakes to their diet as the diurnal raptors. Studies of owl diets are aided by their peculiar tendency of producing pellets—they regurgitate undesirable parts of animals such as hair, bones, and feathers, producing a yucky little blob that they spit out in front of predictable roosts. You can simply gather up the pellets and sort through them to determine what the owls have been eating.

The most accomplished snake killer among American owls is the great horned owl, which takes over night duty in the same open country used by red-tailed hawks during the day. Like their diurnal counterparts, the great horned owl is a generalist predator of a diverse array of mammals, birds, reptiles, amphibians, and invertebrates. They are known predators of 13 kinds of snakes, including formidable opponents like copperheads, prairie rattlesnakes, and eastern diamondbacks. During the day, the southern swamps are hunted by red-shouldered hawks, Mississippi kites, and swallow-tailed kites. At night, barred owls patrol the same territory. This round-faced, brown-eyed owl is also a generalist predator, feeding mostly on small mammals. But their diet includes a large number of ectothermic prey, such as frogs and crayfish. Despite this, the barred owl does not seem to be particularly fond of snakes. Forty-three separate studies show that snakes make up less than 1% of the barred owl's diet.

Most American snakes are harmless tubes of meat. They start out small, grow slowly, and tire easily in a chase. On cool mornings they can't crawl as fast as they can on warm afternoons. Given these disadvantages, it's pretty surprising snakes don't turn up more frequently in the diet of barred owls, not to mention a host of other accomplished predators of the American landscape. They usually make up less than 5% of the diet of carnivorous mammals like coyotes, bobcats, badgers, otters, and raccoons, which otherwise kill thousands of small mammals like mice and voles. They make up a small proportion of the diet of wading birds that otherwise kill thousands of small fish and tadpoles. And although they are captured in large numbers by a few kinds of raptors and other birds, they are practically ignored by the nocturnal owls that hunt the same terrain at night.

I think the reason snakes are not fed upon more often is not because they are more adept at hiding from predators, or that they are only found at low population densities. These factors may play a role, but I have a hard time believing that a barred owl can find dozens of mice and shrews running across the swamp at night but somehow miss the crawling watersnakes. And I have a hard time believing that herons and egrets can patrol the edges of Lake Okeechobee for weeks and only come up with a single black swampsnake. The pattern described in this chapter is that snakes are only killed occasionally by predators, which might indicate they are avoided as prey much of the time.

Despite their small size, snakes seem to be boxing well above their weight class in the food chains of American ecosystems. How they avoid becoming meals is the topic to which we will turn next.

8 · Snake Defense

Several of Nature's People
I know, and they know me—
I feel for them at transcript
Of cordiality

But never met this Fellow
Attended. Or alone
Without a tighter breathing
And Zero at the Bone

EMILY DICKINSON, "A Narrow Fellow in the Grass"

A Snake Propeller

The pasture was planted with hay and dotted with tawny broomsedge rolling out a few acres to the edge of some loblolly pine and sweetgum woods. A narrow row of shrubs lined a storm drain draw that slithered out to a small farm pond next to a disheveled old barn. The Tolchers' was the last house at the edge of the neighborhood, and their backyard bordered the pasture. This was when Atlanta's sprawl began eating up small woodlots and pasturage, and we lived on the frontier of rapidly growing suburbs consuming farmlands. Before then it was small farms and woodlots regenerating from clearing during the Great Depression. Before then it was small plantations growing cotton and cash crops, like the one Margaret Mitchell depicted in *Gone with the Wind* (the real Tara—the plantation upon which Mitchell based her book—was 5 miles down the road). Before then it was the frontier between Georgia and Creek Indian land. Before then it was wilderness. To me it still was.

A gartersnake's stripes can confuse predators as it moves rapidly through the grass. *Photograph by Curtis Callaway*

I was walking down the street on one of those milder summer days when I went around with no shoes or shirt after a rain cooled things off. The pavement was hot with a thin, slimy slick of rain radiating off a humid glow. Though far from the desert, the air was thick with the sweet chemical scent of creosote from all the greased and newly wet telephone poles. Once on a day like that I walked well outside my assigned boundaries clear to the next neighborhood for the better half of an afternoon and only returned after cutting my foot wide open on broken glass. I can't remember my parents' fear but I can imagine it—I don't remember getting in trouble for it. The neighborhood parents worried about kidnappers, murderers, and sexual predators—I was a child during the terror wrought by Wayne Williams—but they still let their kids play outside.

That day near the Tolchers' I was in luck: right there next to their mailbox on the driveway was a big fat gartersnake. Up to that point I had little experience catching snakes and was still a little reluctant to handle them. After all, it is unpleasant to be bitten by anything, even if it's harmless. So I picked it up, but not by its midsection, where I could keep its center of gravity balanced, but by its tail. It immediately reached back around itself to bite me, and became rather ornery rather quickly. Now, I should mention two things about snakes that you might not have known until now. The first thing is that snakes hate to be picked up by their tail: you wouldn't want to be picked up by your toes and dangled in midair. So it is with snakes. The other thing is that—despite their reputation for being harmless and even gentle—gartersnakes are in fact mean as hell.

The snake bolted back and forth in my hand in short waves and looped itself back toward my fingers with its mouth open wide. I got one of my first doses of the pungent, disgusting odor of a snake's musk when it sprayed an oily yellowish paste in a sputtered row on the driveway and down my leg. Still, I held on, wanting to catch this snake and bring it back to my house to put in an aquarium. It lunged and bolted, but I held on tight to the last two inches of the snake's tail. Then the snake did something strange, which I would never tell anyone else about until now.

The gartersnake lunged toward me in a sideways arc as if trying to bite me, then continued going in a counterclockwise fashion, whipping around all the way in a circle until it came back toward my body again. I leaned my arm out farther so it couldn't bite me. The snake orbited again, then again, and then things got really weird. I reached farther and farther out, holding the snake away from my torso as the snake's rotation flattened out more and more. Soon the snake was making its way around in a circle so fast that I couldn't even make out its body by virtue of the blur. I swear that it made a whooshing sound as it turned around and around like a propeller in my hand.

I was scared. Then the snake dropped with a slap to the ground and I had only the tip of its tail in my hand. The stump was trailing blood as the snake made its way into the Tolchers' lawn. I dropped the bloody tail in disgust and was too afraid to chase the snake and try to catch it a better way. I looked around, wondering if anybody else had seen the bizarre display. I walked home embarrassed and confused. Later, I would question whether it even happened at all. Or if it was one of those strange childhood memories that get mixed in your brain with dreams.

I never told anybody about this incident because I was sure they wouldn't believe me. It is told that snakes do all kinds of bizarre and wonderful things when you encounter them, and many of the more fanciful tales are just that —stories and misconceptions passed down from person to person in something between pure cultural mythology and an urban legend. Take the milksnake—the 3-foot-long rodent killer from the eastern United States that gets its common name from the mistaken belief that it steals milk directly from the udder of dairy cows. This story is so prevalent that many snakes in Central America are believed to do the same thing. Of course, no snake does any such thing, and the most frequently offered explanation for the myth is that milksnakes spend a lot of time in old dairy barns because such structures are excellent homes for their rodent prey.

The "hoop snake" legend may trace its ancestry to antiquity and has been applied in North America to coachwhips, mudsnakes, and even glass lizards. As the story goes, to escape an attacker, the snake bites its own tail, fashions itself into a kind of living Hula-Hoop, and quickly rolls down the first available hill.

I had heard of such legends when I was a boy, and I knew they were bogus. And when I would read about the legends, or when I saw nature show hosts talking about them, the stories were always retold with an air of scorn and smug condescension. Who could believe such ridiculous stories? What kind of moron would claim to have seen a snake roll down a hill? Naturally, I did not want to be one of these morons going around telling people I'd caught a snake that turned itself into a propeller to get away. I certainly wouldn't go around telling fellow biologists about it.

Fortunately, years later, I am finally able to confirm that this dubious observation is an actual phenomenon. The late great Henry Fitch, the venerable old master of field biology and author of hundreds of papers on the biology of snakes, wrote a paper mentioning this antipredatory behavior of the gartersnake. It turns out that Fitch witnessed gartersnakes whip themselves into propellers to twist their own tails off. And if Henry Fitch was willing to risk his reputation in a scientific journal, I am finally safe to admit I've seen it myself.

If gartersnakes really do turn themselves into propellers, we can now speculate about whether and how such a behavior could be helpful to the snake. If this maneuver can frighten a budding young naturalist—enough to keep it a secret for decades—imagine what it might do to the psyche of a hawk. Either way, if the snake is able to twist itself away from a predator, leaving only its rather unnecessary tail in a hawk's talons, the snake lives to slither another day.

Henry Fitch
Radio-Tracking Rattlers at Age 93

Snake researchers often follow a standard formula: find a way to catch snakes, measure them, mark them, and then let them go. Catch them again and repeat ad nauseam. From this template you can learn quite a lot about snakes: their population densities, growth rates, and more. The scientists I've chosen to highlight in this book have used novel approaches to study snakes; in some ways, they have gone against the grain of the typical studies used by snake biologists. But the standard approach has its uses, and it was invented and mastered by Henry Fitch.

Fitch began studying gartersnakes in California and eventually landed at the University of Kansas, where for decades he studied tallgrass prairie snakes. He and his students published

Alice Echelle

Henry Fitch radio-tracking timber rattlers at age 93.

important monographs on snakes after studying them with unparalleled thoroughness. Common gartersnakes, copperheads, and western hog-nosed snakes were each treated in detail, and these papers remain standard references. Fitch was still actively gathering data in his eighties when one day he didn't return home from a survey. He'd fallen in a ravine and broke his leg. They found him the next morning, and he was back out in the field within weeks. Near the end of his career, Fitch produced a final, remarkable monograph: *A Kansas Snake Community: Composition and Changes over 50 Years*. This landmark work is the only long-term study of its kind on snakes. After producing something so astonishing you'd think Henry would take it easy and perhaps retire. But he continued radio-tracking timber rattlesnakes well into his nineties with the help assistants and a golf cart, and gave a presentation about this work at the annual herpetology conference in 2002. The meeting organizers made sure to reserve the giant ballroom for his talk, but there were still people standing in the back and squatting in the aisles. I tracked him down at lunch, shook his hand, and had him autograph a copy of his classic *Autecology of the Copperhead*. His eyes were bright and he smiled wide. That book is behind me on my office shelf and it's one of my treasured possessions.

Henry finally passed away a few years later, in 2009, at the age of 99. But his methods and inspiration live on. I was proud to be involved in a paper that detailed the natural history of the southern hog-nosed snake, which we quite consciously modeled after Henry's old monographs. And we pay tribute to Henry every time we check a funnel trap, clip a scale to mark a snake, or describe the habits and general biology of poorly studied snake species.

No legends are necessary to make the defensive behaviors of snakes seem more interesting. The only behaviors that snakes make readily available for observation is their stereotyped and fascinating repertoire of antipredatory behaviors: bluffs, threats, rearings, lunges, shakes, rattles, bounds, barrel rolls, and even playing "opossum." It's a wonder that scientists don't spend more time studying these interesting behaviors, but more research has been done on topics that are far more difficult to observe in snakes: things like courtship and predatory behavior. The great thing about the escape behaviors of snakes is that they are often displayed under certain circumstances by many individual snakes, allowing repeated observations. And usually the snakes readily perform these behaviors in the presence of a human investigator because they consider people a threat.

The defensive behaviors of snakes often follow a predictable, escalating sequence, which was long ago described well by Clifford Pope: "Snakes are first cowards, then bluffers, and last of all warriors." This is quite accurate, and there is now scientific data to support it. But I would quibble with Pope about the first line of defense for snakes. Rather than cowardly escaping—which nearly all snakes attempt after first being discovered—snakes are well equipped at hiding and would overwhelmingly prefer to avoid any encounter with a human to begin with.

Concealment

From picture books—even the one you're now holding in your hands—you may have the impression that snakes are brightly colored. This is not true. Some snakes are colorful, to be sure, but the impression that snakes are boldly patterned with bright colors has more to do with our own preference for striking color patterns than reality. Editors choose visually striking and appealing photographs to illustrate books about snakes, but they are trying to lure you in to buy them. (Just look at the gorgeous snake on the cover of my book!) But snakes love to hide, and the last thing they want is to attract you.

Most American snakes have boring and drab color patterns—Earth tones of gray, brown, and beige—and they are either unicolored, blotched, or banded. This includes most colubrids, viperids, as well as the primitive blindsnakes and boas. Prime examples include common and widespread species whose names suit their often nondescript colors: brownsnakes, earthsnakes, and groundsnakes. Others are similarly plain or brown-blotched snakes: gophersnakes, glossy snakes, prairie rattlesnakes, watersnakes, and hog-nosed snakes. Their subdued colors allow them to blend in well in a variety of backgrounds, ranging from plain dirt and leaf litter to dried grasses and sticks. Their banded patterns further conceal them by breaking up the outline of their bodies; bands and blotches play tricks on the eyes of predators and help the snakes disintegrate into the background.

A copperhead's pattern is among the best camouflage in nature. *Photograph by John Hewlett*

More elaborate chevron or diamond patterns are also common, and these generally help camouflage them even better: eastern diamondback rattlesnakes appear to be boldly patterned when viewed from the air-conditioned side of a zoo exhibit, but in the humid interdune meadows of a South Carolina barrier island, even a five-footer is practically invisible. Similarly, the oranges and subtle pinks of a copperhead's bands may seem vibrant when photographed on a bright-green background of mosses, but on the leafy floor of a Pennsylvania oak hickory forest, the pattern conceals them seamlessly. The khaki gray or even pink ground color of a rock rattlesnake may appear quite attractive when viewed in the hobbyists' terrarium, but these snakes are only as boldly colored as their geological namesake, and not a bit more. The rocky talus slopes of Texas, New Mexico, and Arizona are just as gray, and just as pink. And the snakes match them perfectly. But none of these snakes should be considered brightly colored.

The striped pattern of gartersnakes, ribbonsnakes, whipsnakes, patch-nosed snakes, lined snakes, and crayfish snakes is overlain on a generally dull brown, tan, or olive color scheme. Stripes allow snakes to blend in well in grassy habitats, and have the added advantage of confusing predators when the snake is moving. This so-called flicker-fusion makes striped and banded snakes appear to move faster than they are actually going, so that predators trying to lunge toward the head of snakes might come up with the tail instead. Many snakes that prefer open, grassy habitats during the day are therefore striped.

Several kinds of American snakes are uniformly dark colored or black, which is probably not the best way to disguise a snake but instead possibly a good compromise. Black snakes disappear surprisingly well in dappled sunlight and can easily be mistaken for a thin forest shadow. Many of these are large and active snakes that hunt during the daytime. Examples include black ratsnakes, indigo snakes, black racers, coachwhips, and some rattlesnakes that are sometimes very dark or completely black. Their dark skin also allows them to warm up quicker because it absorbs more solar radiation. It's just like wearing a black T-shirt on a cool sunny day.

There are three greensnakes in America, and I have to admit they are as green as a parrot. These include the rough and smooth greensnakes of the eastern United States, and the rarer green ratsnake of the Arizona borderlands. All three live among the green foliage of trees or shrubs, and they are perfectly camouflaged by grasses, leaves, and vines. Even though they have bright colors, those colors still serve to disguise them.

The drab color patterns of snakes are reinforced by their behavior. The most underappreciated behavior typical of snakes is freezing stock-still. As we learned in chapter 3, snakes are lazy and spend most of their time resting under cover. But even when out in the open, snakes spend long hours lying still. This is most true of ambush hunters like pitvipers. When they detect the approach

A timber rattlesnake's chevrons can conceal it well among leaves on the forest floor. This is a mated pair. *Photograph by John Hewlett*

A rough greensnake spends most of its time in trees or shrubs, where its color provides excellent camouflage. *Photograph by Todd Pierson*

of predators, most snakes will simply lie still and wait for them to pass by. You can observe this almost any time you see a snake. If you see a snake coiled among rocks, simply walk by and see what it does. Nine times out of ten it will just sit there. If it is crawling along and doesn't notice you, you can watch it move slowly and confidently through the environment. But they are wary creatures, and eventually they see you and freeze. If you approach, they then typically attempt to escape by crawling away rapidly. But rapid movements catch the eye of predators like cats and raptors, and snakes know better than to make themselves more obvious. So, if they can, they lie still, confident in their camouflage, almost as if they are watching you pass from the corner of their eye.

This camouflage is so exceptional that even humans need some training before they can find a snake. Predatory animals like hawks, blue jays, and people actively form in their mind a picture of what is known as a "search image" of their prey. This helps them to lock onto their targets while searching for them. Birds of prey often concentrate on one type of prey item so long as it is common, nutritious, and easy to catch. They form a search image for it, and each time they find their prey, they can refine the search image and improve at finding it. People hunt the same way, and it is amazing how hard it is to find a snake without first having a search image for it. Have you ever been in the grocery store looking for an ingredient you've never used before? It's much harder to find something when you don't know what it looks like, isn't it? I never found queensnakes in the swamps where I did most of my snake research until I found one killed on a nearby road. Then I knew what to look for. Within a week, I noticed one basking in a shrub along the river bank I'd paddled by nearly a hundred times. Whit Gibbons told me a story about the day he and some students were paddling the edge of the Savannah River. After seeing nothing for hours, they noticed a brown watersnake and then began counting more, finding about 30 as they drifted down the river a few hundred yards. Then they noticed a rough greensnake stretched among the vines, briers, and low-hanging strainers. They paddled back upstream from whence they'd just came, and counted 30 greensnakes they hadn't noticed on their first pass. The camouflage of snakes is no doubt geared toward defeating a predator's search image, and it's uncanny how well they can hide themselves. Sometimes you need to know they are there before you can find them.

There are only a few American snakes with bold colors, and they flagrantly advertise their presence with them. These include the red-, yellow-, and black-banded coralsnakes and their mimics. Mimicry is when a harmless species develops a feature that makes it look like a venomous one. Nature is full of wondrous and sometimes improbable examples of mimicry; the monarch and viceroy butterflies are perhaps the most famous example in North America. Monarchs feed on milkweeds, making their flesh distasteful. Because birds cannot palate the monarch, the edible viceroy—which looks just like the monarch—is spared. For some time, scientists thought that perhaps bright colors serve to break up the coralsnake's pattern or to disguise the snake at night; in dim light, reds and yellows rapidly wash out. But it is now thought with some confidence that coralsnakes use their bright colors to warn predators, and several kinds of nonvenomous snakes take advantage of this characteristic, becoming dead ringers for the dangerously venomous coralsnakes. These include the scarletsnake, short-tailed snake, and scarlet kingsnake of the Southeast, whose distributions overlap with the eastern coralsnake. In Texas, the scarletsnake, milksnake, and long-nosed snake have distributions that overlap with the Texas coralsnake. And in the Sonoran Desert, the Arizona coralsnake shares its range with sandsnakes, shovel-nosed snakes, and groundsnakes that look much like it. Such gorgeous snakes as the Arizona and California mountain kingsnake are brightly colored with red, white, and black bands and possibly enjoy some protection from predators owing to their similarity of appearance to coralsnakes.

Some gartersnakes also have bright-red colors that advertise their toxicity to predators; in this case, the gartersnakes are actually poisonous because they eat newts and accumulate their toxins in their tissues (see chap. 6).

Coralsnakes advertise their toxicity to potential predators, and they are mimicked with some skill by nonvenomous colubrids. *A*, Harlequin coralsnake *(photograph by Dirk Stevenson)*; *B*, Sonoran coralsnake *(photograph by Pierson Hill)*; *C*, scarlet kingsnakes *(photograph by Dirk Stevenson)*; *D*, red milksnake *(photograph by J. D. Willson)*; *E*, Sierra mountain kingsnake *(photograph by Marisa Ishimatsu)*.

Bluffing

American snakes incorporate a range of delightful and bizarre behaviors into their antipredatory repertoire to avoid ever having to fight with another animal. Many of these behaviors are poorly studied and poorly understood. We are not entirely sure what, if anything, many of them do to discourage predators. That is one of the most exciting things about snake behaviors: they are fascinating, yet sometimes their significance is still waiting to be discovered.

Principal among these is the universal tendency for snakes to squirt a foul-smelling musk from a unique pair of glands at the base of their tail. These glands can spray a pair of fine, well-aimed mists of stink toward their enemies. You can see the glands evert and spray if you quickly turn a snake over when you first capture it. In addition to the faint musk produced by these glands, most snakes will also let fly a large amount of urates from their cloaca when handled or attacked. This creates a perfect mess of wet smells and toxic waste that typically end up on your clothes. The smells produced by the anal glands are characteristic of the snakes that make them. Gartersnakes are

C

D

E

Sara Ruane
Revealing the Mysteries of Milksnakes

Sandya Viswanathan

Cemeteries—especially old, neglected ones—can be sanctuaries for wildlife and are sometimes visited by people looking for suitable substitutes for parks. St. Catherine's is divided into two halves by woods and a hemlock-lined creek, each half-perched atop ridges overlooking Moscow, Pennsylvania. In autumn, neatly planted oaks lining the cemetery lanes are ablaze in orange, scarlet, and yellow. But in the summer—that glorious Pennsylvania summer you survive the long, sooty winter to deserve—the trees shimmer with a nearly electrical green.

One summer day, a little girl found a pile of old flowers behind the cemetery toolshed. She had learned from her grandmother that by digging through a compost pile, little girls can find treasures, like bright-red centipedes and slimy worms. She dug through the layers and found an object so beautiful, so fascinatingly alive, that her whole life changed that afternoon in northeastern Pennsylvania. There, under a small green plastic pot wrapped in silvery foil, was a 10-inch snake the width of a pencil, ringed in red, black, and white. Filled with incurable excitement, she ran over to share her discovery with her beloved grandmother.

This and other childhood encounters with snakes cemented a fascination with serpents so intense that when she was in the fifth grade, Sara Ruane told her teacher, "I'm going to be a herpetologist." Although she enjoys nature generally and finds other creatures fascinating, to this day she says, "All the other stuff is good, but a snake is special. They are harder, more elusive. Catching a snake is a big event." Sara used cutting-edge genetic techniques to unravel the relationships of some of the most stunningly beautiful snakes: the American milksnakes. These snakes come in a variety of striking color patterns, with body rings in combinations of red, yellow, white, and black. Many of these bold color patterns enable milksnakes to mimic venomous coralsnakes and thereby avoid predation. The snake Sara found at the cemetery when she was 7 years old—the one that started it all—was a baby eastern milksnake.

Sara's studies demonstrate that varieties of milksnakes are best treated as separate species. This realization has been a long time coming; I've always had a problem considering the scarlet kingsnake the same species as the eastern milksnake, which is a much larger species with much more somber adult colors. Thanks to Sara, there are now many more recognized species of American milksnakes. She continues her research on the evolution of snakes, and this work has now taken her to Madagascar, where her work ethic inspired her former advisor, Frank Burbrink, to describe her as "the most hardcore person I've ever been in the field with."

Sara dedicated her PhD dissertation on milksnakes to her late grandmother Catherine Ruane, who died shortly before she completed it. The dedication reads: "A special thanks to Grama Bell, Catherine F. Ruane, who never failed to encourage my love of nature and of snakes in particular."

Grama Bell is buried at St. Catherine's cemetery in Moscow, Pennsylvania.

quite smelly and often paint their targets with an oily, creamy lather of urates. Cottonmouths have a sickly sweet-smelling musk with hints of nutmeg. It is almost—but not quite—pleasing to the nose. The musk of timber rattlesnakes has been likened to the masculine scent of an old billy goat. Ringneck snakes have a surprisingly potent stench for such a small snake; it is a cheesy smell, with nutty notes. By far our worst-smelling snake is the diamondback watersnake, which matches its anal gland's potency with quantity, producing a wretched aroma accompanied by massive production of urates. You cannot catch one of these snakes without receiving a hearty smattering of thick, yellowish spuzz. It smells just like a skunk.

Every herpetologist has intimate experience with these smells, but none of us have much of a clue what purpose they serve. The smells are awful but nowhere near as powerful as skunks or even weasels; they have never deterred any of us from our work. But it is possible that a well-aimed shot of musk may thwart predators with keen senses of smell. Coyotes, cats, and dogs are not thwarted by these repellants, but it is possible that bears are.

Another uncouth behavior involving the snake's rear end is "cloacal popping"—when captured, snakes sometimes rapidly and rigorously turn their cloaca inside-out. This can supposedly be done with enough enthusiasm to make a barely audible popping sound. To me it sounds more like a sickening squishing. Examples of snakes with this remarkable capability are the Arizona coralsnake, and the tiny and poorly known Chihuahuan hook-nosed snake. It is unknown what, if anything, this does to thwart predators, but it has been suggested that it perhaps baffles and confuses them. Before determining the purpose of cloacal popping, it would be difficult to rule out the possibility that the snake is simply so scared that it loses control over its continence. But it is strange that two uncommon, burrowing species of the southwestern deserts share the same bizarre behavior.

Many snakes curl themselves into a small ball and hide their head under their coils when attacked. This behavior is best known in the rubber boa, which hides its head and presents its blunt tail as a false head in what is referred to as "automimicry." In this way they can receive several blows from potential predators while their real head looks for an escape route. Many rubber boas have severely scarred tails that are a testament to their previous escapes. Coralsnakes also hide their heads in a similar fashion, but this deadly snake can quickly lunge out from its hiding place to inflict a serious bite. The only coralsnake I've tried to catch wriggled forward and backward with equal dexterity, and it was difficult to tell which end was which. Even rattlesnakes hide their heads as a last resort when they are under attack from kingsnakes, and some do so when approached by people.

Hiding the head is one way to protect it, but snakes have another way to distract attention away from this most vital feature of their anatomy. Most snakes rattle their tail when threatened. This leads many people to think

A New Mexico milksnake hides its head among its coils.
Photograph by Crystal Kelehear

that just about any snake is a rattlesnake. Ratsnakes, gophersnakes, cottonmouths, copperheads, and many other species do it, and when rattling among dry leaf litter, it sounds reminiscent of a rattlesnake. To know if it's a rattlesnake, you have to see if it has the knobby, sound-producing structure that gives rattlesnakes their name. That other snakes shake their tail is not thought to be mimicry of rattlesnakes, since tail vibration and wagging displays are common and exhibited by snakes all over the world, even in places where there are no rattlesnakes. Instead, this behavior was surely present first

A rosy boa hides its head in its coils and presents its tail to attackers. *Photograph by Bob Hansen*

in many snakes, and the fully formed rattle became an even more effective deterrent to predators. Tail displaying by nonvenomous snakes draws the attention of predators away from the head toward the tail. If the tail is attacked instead of the head, the snake will possibly get away unharmed.

Tail-rattling behavior usually happens when snakes are approached closely and have no chance to escape. It is accompanied in some snakes by additional defensive behaviors that serve to frighten their enemies or make them more difficult to swallow. Most snakes inhale air and inflate the large sacular section of their lungs, quickly becoming twice their starting girth. By increasing their apparent size in this way, they may thwart smaller predators that were already on the fence about whether the snake was small enough to handle. Many predators also feed on prey they are certain they can fit in their mouths—so called gape-limited hunters—so larger prey that are too big to swallow are released.

This ballooning behavior may frequently save smaller snakes and limit their potential predators.

Many snakes also rear back into defensive postures when threatened, piling up the first half of their bodies into coils with their faces toward their attacker. These greatly resemble the defensive postures of rattlesnakes and may serve as a form of mimicry, but such threat displays alone may be enough to deter small predators that must face towering coils bending menacingly toward them, instead of what until seconds ago was a small, thumb-wide snake. Pinesnakes and gophersnakes rear in this fashion with their mouths open, hissing deeply, and suddenly lunge at their enemy. While lunging, their hiss comes to a startling crescendo. This must surely frighten their enemies; it's even difficult for me to stand my ground when they lunge like this.

Pinesnakes are among the few that actually hiss. They have a special flap of skin in front of their windpipe around which air races to produce the sound. Some other snakes can create a slight hiss when they rapidly expel air from their lungs, but the notion that snakes spend most of their day crawling around hissing is largely derived from cartoons.

In addition to these rather straightforward defensive behaviors, there is a litany of additional displays exhibited by American snakes whose function is far from clear. Ringneck snakes turn up the underside of their tail into a curly-Q and present their bright yellow, orange, or red belly colors. This prob-

A red cornsnake raises its front coils in a defensive display. *Photograph by Daniel Wakefield*

Ringneck snake displaying its brightly colored tail. *Photograph by Todd Pierson*

ably serves as a distraction while the snake's more vulnerable head searches for an escape route. Mudsnakes gently dig the sharp tip of their tails into your hands when you catch them; it is not painful but might be disconcerting to some predators. The belly colors of many snakes, including the bright-scarlet checkerboard pattern of mudsnakes, directly contradict my insistence that snake colors are dull and designed for concealment. But why do they have such bold colors on their bellies, where presumably nothing ever sees them unless the snake is being handled? My only guess is that somehow these colors may briefly distract or frighten potential predators. Other animals have small patches of bright colors concealed by some structure and are otherwise camouflaged. When attacked, their "flash colors" are revealed, which distracts or startles visually oriented predators. An example is the underwing moth, which has upper wings that supremely camouflage the moth while resting on tree bark. When disturbed, the moth unfurls these wings to reveal an additional pair of bright-orange wings. Io moths take this a step further—when disturbed, their underwings reveal a large pair of haunting and remarkably convincing eyes. These devices appear to thwart birds that try to attack the resting moths, and it is possible that the bright colors found on snake bellies work the same way.

Red-bellied snakes have colorful bellies and also exhibit what might be the all-time strangest defense behavior when captured. It also happens to be one of my favorites. One side of the upper lip curls to reveal the upper jaw and its teeth—an Elvis sneer. Anybody's best guess is that they are threatening with bared teeth just like a dog, but what I can't figure out is what predator would be intimidated by this tiny, harmless snake.

The most complicated defensive behaviors are performed by the hog-nosed snake, a medium-sized, thick-bodied snake that occurs throughout the eastern United States. It comes in several color patterns, ranging from an attractive reddish orange with a checkerboard of brown blotches, to solid olive or black. When approached, they sometimes erupt into a riotous combination of behaviors—rattling their tails, coiling the rear of their tail into a curly-Q, flattening out their head like a cobra, gaping their mouths wide, and rearing up the first part of their body several inches from the ground. They often lunge repeatedly at their foe. When pressed further, the snake then performs its most famous act: playing dead. They turn over on their back, their mouths wide open, tongue hanging out. If they have eaten recently, they slowly regurgitate their prey and with great amounts of saliva. They sometimes evert their cloaca so that it looks like their guts are hanging out. If turned back onto their belly, they quickly unfurl onto their backs and continue the performance. My buddy Dave Steen swears that if you walk around while they are doing this, the snake's gaze follows you "to see if you're buying it."

The earliest naturalists noted the antics of the hog-nosed snake, yet decades later we are no closer to understanding why they do it. The most frequent throwaway answer is that for some reason predators lose interest in them when they play dead. Perhaps bobcats and other felines drop them if they don't move; watching my own cats quickly lose interest in inanimate objects makes this seem plausible. But it seems to me that a determined predator like a coyote would simply rejoice and happily commence feeding if their quarry died instantly after capture. Others have suggested that adult raptors quickly kill prey and return to their nest, so a quickly killed yet surreptitiously alive hog-nosed snake might be dropped off at the foot of nestling hawks. The snake could then conceivably slither out of the tree to safety. This scenario is certainly feasible. And there is at least one observation of an apparently dead gartersnake that was dropped off at a nest, regained consciousness, and escaped. But confirmation that hog-nosed snakes frequently avoid predation this way is lacking.

Venomous snakes also bluff. The cottonmouth is most famous for its mouth-gaping behavior, which is where they get the name; the inner linings of their pink-white mouth are presented to approaching enemies. Often the first sign of a cottonmouth's presence is a quick flash of their open mouth. The message of this warning is fairly obvious: "if you come closer and mess with me, I will bite you." The behavior is reflexive and one of the few antipredatory responses that have been studied in some detail. Yet we know little about the responses of wild predators to this display.

Fortunately, one fascinating interaction was caught on tape. Paul Andreadis, a dedicated student of cottonmouth behavior, has spent long hours observing cottonmouths in swamps all over the South. His research program is simple: he goes out, finds a cottonmouth, and uses a video camera with infrared func-

A

B

The defensive repertoire of hog-nosed snakes. *A*, hog-nosed snakes often begin their antics by spreading their neck in a cobra-like threat display *(photograph by John Hewlett)*; *B and C*, they then coil their tail while lunging with their mouths open *(photograph by the author)*; *D*, when this fails, they resort to their most famous act: playing dead *(photograph by Todd Pierson)*.

C

D

tion to watch it all night. Most of his movies are unbearably boring; the cottonmouth just sits there all night in its ambush pose doing nothing. But in one remarkable sequence, a raccoon ambles along toward one of Paul's cottonmouths, approaches to within an inch of its flat head, and sniffs it. To see what happens next, you need to watch in slow motion. The cottonmouth's jaws spring open, and the whites of the raccoon's eyes emerge as it doubles over backward and runs away. It seems like raccoons understand the message.

Rattlesnakes have one of the most famous warning devices in the animal kingdom, and evidence suggests that their enemies get the message. The buzzing sound of their rattle is caused by special overlapping hard scales of the tail; a hardened scale at the end of the tail acts like an anchor and clings to the outer covering when the snake sheds. Each time the snake sheds, a new segment is caught and added to an ever-growing rattle. The rattle segments are hard, hollow, and overlap intricately into a segmented organ; if you pick up a rattle that has been hacked off a live snake and shake it, it makes a nice, dry maraca sound. But attached to the tail of a live rattler, it gives an unmistakable loud buzzing.

In my experience, most rattlesnakes do not habitually give away their presence with the rattle. I have had few encounters with rattlesnakes like those you see at the movies: the cowboy is walking along, hears the rattlesnake, and

A cottonmouth displaying its characteristic open-mouth gaping behavior. *Photograph by Noah Fields*

A brown watersnake exhibits gaping behavior, perhaps to mimic a cottonmouth. *Photograph by Noah Fields*

A brown vinesnake displaying its open-mouth gaping behavior. *Photograph by Thomas R. Jones*

A western diamondback rattlesnake exhibiting its unforgettable defensive tactics.
Photograph by Curtis Callaway

then realizes it is there. Rattlesnakes have been referred to as "gentlemanly" because they give fair warning, and this is often the case when you approach and even try to handle them. But I have only walked up on three rattlesnakes that alerted me to their presence by rattling. Most times rattlesnakes would rather let you stroll on by without alerting you.

In addition to rattling and gaping, most venomous snakes will assume a characteristic defense posture when threatened. Rattlesnakes pile up a few coils deep and continually face their attacker with the rattle typically buzzing just behind or to the side of the coils. If you come within range (typically about a foot away or less) they strike rapidly, often hissing simultaneously, although many of these are bluff strikes extended with the mouth closed. Often they begin probing for an escape route with their tail while their head and neck stay locked on you.

The threats, postures, shapes, and sounds of our venomous snakes are characteristic and undoubtedly quickly learned or instinctually avoided by predators. It could therefore be quite an advantage to a nonvenomous snake to be mistaken for a venomous one. The most obvious examples are the coralsnakes and their mimics. Coralsnakes advertise their venom with bright colors; in this way they are trying to quickly educate the local predators about the danger they pose. It does them no good if they are attacked by a predator who won't learn from the bad experience it gets from being bitten—or from a predator who won't teach its young to avoid more coralsnakes. Therefore bright colors reinforce the bad experience. Coralsnakes and their mimics share the same habitats throughout the Americas; where black-and-red coralsnakes live, there are black-and-red mimics; where black-yellow-red coralsnakes live, there are black-yellow-red mimics. This pattern would not be expected to arise by chance or any other mechanism besides mimicry. There is also evidence that raptors and other predators attack coralsnakes and their mimics less frequently than plain-colored snakes.

The degree to which nonvenomous snakes mimic pitvipers is less certain. Many nonvenomous snakes flatten their heads greatly when threatened, which combined with ballooning behavior can make normally skinny snakes look remarkably like pitvipers. Pitvipers generally have a heavy-bodied appearance and a distinct neck—the so-called "triangle-shaped" or "diamond-shaped" head that many people think is an infallible way to identify a snake as venomous. Although it is true that pitvipers have this feature, nonvenomous snakes can appear this way with a little imagination, and many can easily flatten their heads as well. There is now evidence that this behavior is a form of mimicry; a study in Europe determined that local raptors attacked clay models of snakes with narrow heads more frequently than those with triangular heads. Raptors may actively avoid snakes shaped like pitvipers, and instead choose harmless snakes with rounded heads. Hog-nosed snakes, watersnakes, ratsnakes, cornsnakes, and glossy snakes are all known to flatten their heads when approached; this behavior is likely intended to fool would-be predators into thinking they are venomous. Given the number of people who also misidentify these snakes as venomous, I would say that it's working.

The color patterns of many nonvenomous snakes are also suspiciously similar to pitvipers that they share their habitat with. The southern hog-nosed snake is reddish tan or dusky with brown square checkerboard blotches, and looks remarkably like the pigmy rattlesnake it shares its sandy pine habitat with. Western hog-nosed snakes, Great Plains ratsnakes, fox snakes, and massasaugas all live in the same grasslands, and their blotched patterns—including ornate head markings—all look remarkably similar. Gray-banded kingsnakes are handsome snakes that come in a variety of patterns. They can be gray with black bands, and they occupy rocky habitats alongside similar-looking and similarly variable rock rattlesnakes. All of these lookalikes have been proposed as possible examples of mimics but have not been confirmed with experimental or corroborative evidence.

Rattlesnakes and gophersnakes throughout the western states look very much alike, and there is some evidence that gophersnakes are actively mimicking rattlers. The similarity doesn't just stop at the color pattern; when attacked, gophersnakes rear back in a piled coil quite similar to that of rattlesnakes, and their loud hissing is similar to the buzzing of a rattle. But their color pattern similarity is mostly due to the habitats the two snakes live in; in grassy areas, both snakes converge on a similar pattern that helps camouflage them. In brushy areas, both species are slightly different. In this case, the two snakes appear to converge on a similar pattern because it serves to conceal them. Any additional benefit the gophersnake receives from looking like a rattlesnake is probably incidental.

Nonvenomous watersnakes occasionally mimic the gaping behavior of cottonmouths, although this occurs so infrequently that the phenomenon has never been systematically studied. I have only seen watersnakes do it a few times—a few times by brown watersnakes and once each by a northern watersnake and banded watersnake. The experience with the latter snake was quite memorable. A few years ago, I was walking along the edge of a small gum pond in southern Georgia when a large, dark snake suddenly snapped to attention inches from my feet. Out of the corner of my eye, I saw a stout snake and a flash of white. I jumped to the side and swore loudly in the same second, but quickly recovered; my next thought was that I had a close encounter with a cottonmouth. Instead, I was surprised to find that it was just a watersnake.

Fight

Even nonvenomous snakes are surprisingly good fighters. There are several well-documented observations of raptors that got more than they bargained for when they attacked large snakes. The earliest of these was seen by William Bartram, the eighteenth-century botanist and naturalist who traveled the Southeast by foot and horseback when the Creek Confederacy and Cherokee Nations still composed most of what is now Georgia and Alabama. Bartram observed a hawk and a coachwhip engaged in mortal combat; the snake wrapped its coils around one of the bird's wings and prevented it from using it for flight, when, "upon coming up, they mutually agreed to separate themselves, each one seeking his own safety, probably considering me as their common enemy." He goes on to correctly assign blame for the contest, writing, "I suppose the hawk had been the aggressor, and fell upon the snake with an intention of making a prey of him; and that the snake dexterously and luckily threw himself in coils round his body, and girded him so close as to save himself from destruction." He then begins a new paragraph with the surprising line: "The coach-whip snake is a beautiful creature."

Several more observations involving raptors, owls, and snakes have been

A gophersnake in a life-or-death battle with a red-tailed hawk. *Photograph by Ace Kvale*

reported since Bartram, and extend all the way to the Internet age; Google "snake attacks bird" and you can watch for yourself. In all of these cases the snake is not attacking the bird, but rather defending itself—sometimes quite skillfully. Racers, coachwhips, striped whipsnakes, gophersnakes, rubber boas, and ratsnakes have successfully grappled with great horned owls, red-tailed hawks, and red-shouldered hawks—the three birds known to feed most frequently on snakes. Usually the snake uses a firm, constricting coil to disable one of the bird's wings, and another coil finds its way around the bird's neck. Hogtied in this fashion, the bird cannot fly away. Instead, it tries to fight back using vicious pecks from its curved beak. Examples of snakes killing the bird are not unknown; a rubber boa was found alive with a red-tailed hawk it had strangled to death. Likewise, cases where both participants were killed during the fight are known; a great horned owl was found dead with a black racer wrapped around its neck. The racer sustained lacerations and succumbed to wounds inflicted by the attacking owl.

Venomous snakes also defend themselves with enthusiasm, but venom is used primarily to incapacitate and digest prey; for defense it is something of a last resort. In my experience there seems to be a point at which venomous snakes switch from their usual defensive repertoire—lying still to escape detection, evasive escape behaviors, bluffs, and bluff strikes—before intentionally striking to inflict a venomous bite. When being handled, cottonmouths and rattlesnakes (I always do this safely with steel tongs; never with my hands) lie still until the tongs touch, and then they try to squirm away. Rarely, certain grumpy rattlesnakes strike the incoming tongs on first approach. But these strikes are usually bluff strikes. Bluff strikes are more like punches—the snake lunges forward rapidly, as quick as a real strike, but with the mouth closed. I can tell the difference between a bluff strike and a real strike because I can feel, and sometimes hear, the ticking of the fangs when they hit cold steel. The bluffs must be an obvious warning to potential predators, a way of saying, "listen . . . you know I'm dangerous, and next time you're going to be sorry." Only when the snake is truly hassled, often after a full minute of handling with the uncomfortable tongs, will it finally and quite purposefully munch down with the full force of its fangs. But this is just my general impression after handling hundreds of cottonmouths and rattlesnakes. There is actually solid scientific evidence that pitvipers defend themselves in a way far more courteous than most people would ever imagine.

This black racer was found dead next to the great horned owl it killed in self-defense. *Photograph by Roger Perry, courtesy of the US Forest Service*

This evidence comes from a remarkable study by Whit Gibbons, who determined the precise sequence of defensive behaviors of cottonmouths toward humans. Probably the most remarkable thing about the study is that nobody had bothered to do one like it before. Whit had the bright idea of walking up to cottonmouths and writing down what they do when you mess with them. He even went through the trouble of stepping on the cottonmouths (lightly, and with snake-proof boots), and picking them up (with an ingenious phony arm). Over half the time, when approached, cottonmouths slither away as fast as they can. Some stay put, as if they are hoping you'll just go away. The open-mouth threat display characteristic of cottonmouths is given about 78% of the time, and even when snakes are stepped on, they bite your boot only 5% of the time. When picked up by a fake arm, the snakes only bit the hand 36% of the time, and this was one of the only things the researchers could do to induce that dreaded response.

Prairie rattlesnakes have similar tendencies. When encountered out in the open alone, they lie still and rely on their camouflage, a behavior referred to with the unfortunately stiff term "procrypsis." When they realize there is no chance of avoiding detection, 44% of the time the rattlesnakes then quickly attempt to escape, often crawling backward rapidly while simultaneously turning their necks and heads to face their antagonist. As they burst into movement, they give a brief, explosive rattle. They then continue by silent running; crawling away stealthily toward cover or some familiar burrow. For a potential predator, this is surely a jarring and baffling experience, and I can personally vouch for its effectiveness. Rather than attempt escape, a small number of prairie rattlers (about 12%) will immediately stand their ground, assuming a tight, cocked coil or piled-up defensive posture, ready to strike. As often as 7% of the time, prairie rattlesnakes will then "assume the fetal position"—burying their heads submissively within their coils like cowards. Only 6% strike as a first resort.

When approached near cover (a shrub or rock outcrop), either in groups or alone, their defensive strategy is different. Prairie rattlers are much more likely to lie still in apparent hopes that they will remain undiscovered. They are then about equally likely to either attempt escape or defend themselves by coiling and rattling. Snakes near cover are apparently far more confident in their surroundings and their likelihood of escape; they are extremely unlikely to strike.

Venomous snakes are therefore reluctant to fight with their fangs. But during a predation event, venomous snakes quickly run out of less aggressive options and use their venom. And there are remarkable observations that show venomous snakes can use venom to turn the tables on would-be predators. The black bear killed by a timber rattlesnake mentioned in chapter 7 is a prime example. A bobcat seen feeding on a western diamondback rattlesnake appeared dazed and stumbled around, succumbing to venom. There are also several reported instances in which predatory birds were found dead next to the snake that killed them. These include red-tailed hawks killed by a cottonmouth and eastern diamondback rattlesnake. Coralsnakes attacked by a loggerhead shrike, American kestrel, and red-shouldered hawk survived the attack by biting their attackers, which then released the snakes. In another case a red-tailed hawk found feeding on an eastern coralsnake became progressively uncoordinated, and after an hour it fell flat, dead.

Several American snakes show special behaviors that may help them survive close encounters with predatory snakes. Small snakes will tie themselves into knots in an attempt to thwart consumption by other snakes. In this way they make themselves much bigger and much harder to swallow. A smooth earthsnake was found knotted in the jaws of a black racer that was having a hard time swallowing it. The earthsnake was able to squirm away safely when a person came by to see what all the commotion was about. Pitvipers exhibit a special behavior when attacked by kingsnakes. Instead of trying to bite the kingsnake, which would have no effect, rattlesnakes and copperheads align themselves with their body facing toward their enemy, raise off the substrate, and fling a solid coil against the kingsnake's snout. This movement, which is referred to as "body bridging," has been likened to a human throwing an elbow. A good solid strike on the head might disorient the kingsnake, perhaps especially inexperienced ones. This causes the kingsnake to hesitate, which may be enough of window for the rattlesnake to escape.

Probably the most common outcome between a snake and a larger predator is that the predator kills and eats the snake. But rare glimpses into the battles that occur between snakes and predators show that the outcomes of such contests are far from certain. Large rattlesnakes may have little to fear from other animals and can therefore be considered apex predators. And when it comes to defending themselves from humans—after many attempts to conceal themselves, escape, or bluff their way out of a fight—snakes will at last turn to venom. Predatory strikes and defensive strikes are different and can involve different volumes of venom; snakes can control the amount of venom injected into their target. Often this means that even if you are bitten by a venomous snake, there is a good chance you will not be envenomated and perhaps you will not even need treatment.

But the danger of venomous snakes for humans should never be underestimated. These dangerous snakes, and their relationship with people, are considered next.

9 · Dangerous Snakes

After people and the animals were created, they all lived together. Rattlesnake was there, and was called Soft Child because he was so soft in his motions . . . Elder Brother pulled a hair from his own lip, cut it to pieces, and made it into teeth for Soft Child.

"If anyone bothers you," he said, "bite him."

Pima tradition.

KATHARINE JUDSON, *Myths and Legends of California and the Old Southwest*

A Close Call

I took my parents on a hike out to Bear Meadows—a spectacular Ice Age spruce bog marooned in the crooked ridges of central Pennsylvania. A 4-mile trail encircles the bog under tamarack, black spruce, and through thick carpets of sphagnum moss. Such habitats are far more common 400 miles north in Canada. Swamp sparrows and northern waterthrushes breed during May, along with a diverse assortment of warblers I used to see only when they were en route to dark northern coniferous forests: black-throated greens, black-throated blues, Canadas, and many more. The thickets of rhododendron are home to the bog's namesake, though I never saw one there. We walked the trail, and I tried to show my parents the wonderful diversity, stopping to try and "spish out" hooded and blackburnian warblers. My parents, always astonishingly inept with binoculars, saw nothing. I switched gears, trying to bring them some kind of frog or anything I could get my hands on. Of course, they could not have cared less about the wildlife and were just glad to go on a hike with me. Still, manically, I wanted them to see something exciting.

I got my wish when I saw a big black timber rattler a few feet ahead of my mom. It was stretched out in the dappled sunlight of a trail opening. Her path would have taken her right past it. I stopped her with my arm and told her to be calm and look. It took her a few seconds to find it. Now, my mother is a deeply religious woman and for some reason I wanted this snake to surprise her. With profound prejudice, I was hoping she would reveal her profound prejudice about the snake and panic, so I could then have the opportunity to smugly show her there was nothing to fear. Instead, she couldn't stop talking about how beautiful it was. Undaunted, and unnecessarily—for my mom could sense that she was in no danger—I demonstrated how safe she was. I walked toward the snake, continuing the path she was walking before I stopped her. I walked right past the snake, and it never moved. It rattled with little enthusiasm when I got a little closer to take its picture. That was it. She would have kept right on walking past it, and she would never have seen it. Which is just what the rattler wanted.

I've had close calls with several venomous snakes. I've stepped on no fewer than six cottonmouths and stepped over several more. I've walked within easy reach of eastern and western diamondbacks on purpose and leaned over an unseen copperhead whose head was within 10 centimeters of my jugular. In all of these cases the snake either remained still or tried to get away. None struck at or bit me.

These encounters had nothing to do with my being a snake biologist. And these do not include some fairly close calls I've had working with venomous snakes in a zoo setting. If you work with venomous snakes in captivity, you're asking for it. And I should graciously and magnanimously admit that if you work with venomous snakes at all, you're asking for it. Herpetologists, most particularly snake biologists, are in fact disproportionately prone to falling victim to their study subjects. This probably comes to many as no surprise, but many snake biologists like to think that if they're careful it won't happen. But it does. I usually tell people it's not a question of "if" but "when." Still, most herpetologists die of old age or cancer just like everyone else. And at the time of this writing, I've never been bitten.

If you want to work with a dangerous animal, look no further than the chimpanzee. The chimpanzee is arguably the ultimate model organism for biology: they are closest genetically and biologically to humans, so by studying them, we can learn the most about ourselves. But there's a reason why lab mice and rats are more common subjects. They won't trick you into getting close to their cage so they can bite your fingers off. There's a saying that you

I almost stepped on this western diamondback rattlesnake hiding in a vegetation clump. The snake never moved while I photographed it.

can tell which primatologist studies chimpanzees because they're the ones missing fingers. The risks posed by venomous snakes and chimpanzees are highest for the scientists who would study these animals. But even for snake biologists, it's a pretty rare event for one of us to be bitten and even rarer to be killed. For the public, the risk is practically nonexistent, even if you're an active outdoorsman.

So, while working with venomous snakes is dangerous, they are not the most dangerous animal you can find to work with. And although some herpetologists have lost some of the sensation in one of their fingers (or an entire finger) to a previous engagement with a pitviper, most have only one such experience in their past, and perhaps several close calls. Most of my close calls with venomous snakes had nothing to do with me working with venomous snakes, except that I was probably more likely to be aware that I had a close call to begin with. Still, this didn't save me from the snakes. Instead, the snakes spared me. If you are a hiker, hunter, camper, fisherman, rock climber, mountain biker, caver, whitewater enthusiast—any kind of outdoorsman or outdoorswoman—you've probably had just as many close calls and just didn't realize it. Rather than being dangerous, there seems to be no limit to the generosity and patience of venomous snakes toward humans.

It is common for snake biologists to defend venomous snakes, and we have undeniable statistics that put the risk of envenomation or death by snakes in perspective. I have my favorites, and some probably sound reasonable and familiar: in the United States you are more likely to die by lightning strike than by a bite from a venomous snake. But here is a more complete list of risks that far outweigh the risk posed by venomous snakes: death by scalding water, death by drowning in a bathtub, death by choking on your own vomit, death by falling, death by dog attack, death by bee sting. My favorite: you are over three times more likely to be put to death by capital punishment than you are to be killed by snakebite. Snakebites are among the rarest causes of death in the United States. Despite the low risk, hundreds of people put themselves at greater risk by trying to kill venomous snakes because they consider them a threat. By this logic, you would be better served trying to kill your neighbor's dog.

But it would be disingenuous to say that venomous snakes pose no risk to humans. To say so is also elitist and ethnocentric. In other parts of the world, snakes are a considerable medical risk. This is in large part because of the presence of a handful of extremely grumpy, highly toxic species whose distribution overlaps with high densities of shoeless humans without access to good hospitals. Antivenom research and manufacture are shamefully underfunded in the developing world.

Although it is not well publicized, the most deadly snakes in the world are not those you often hear about. The Discovery Channel and other pop-science formats perpetuate the myth that the top 10 deadliest snakes are all found in

Jayme Waldron
Rattler Tracker

Jayme was out on a March afternoon in a magnificent stand of longleaf pine—not a forest, really, but what should properly be called a savannah, a grassland with perfect narrow boles of scraggly barked pines with crooked tops. She checked the overwintering locations of her rattlesnakes, noting their use of stump holes and other retreats for hibernation. The breeding season began in July and ended in October, and then the snakes settled in for a short winter under the salt-and-pepper South Carolina sand. She followed the mechanical clicks of her receiver and suddenly realized this would not be a typical encounter with her snake. She peaked over and saw the plump female, with what she described as "literally a stack of snakes on top of her." Two males were on top, and one was copulating with her. Two additional males were piled right next to them. She explained that, without telemetry, she would never have witnessed this observation, which flew in the face of the idea that eastern diamondback rattlesnakes have a discrete mating season in late summer. She says "telemetry allows you to not walk right by them."

Jayme grew up among the endless green hills near Coalton, West Virginia, and soon "became obsessed with salamanders and frogs. My parents never taught me to be afraid of those things."

Amy Franklin

Soon she added ringneck snakes, brownsnakes, and wormsnakes. She favors snakes because, like many snake biologists I spoke with, she "cheers for the underdog." She can't understand why, despite their shy and retiring nature, "Everyone hates them. Everyone thinks they're ugly, and slimy. But they're not ugly. They're beautiful."

For her PhD research, Jayme followed 21 beautiful eastern diamondbacks and 18 canebrake rattlesnakes for four years, examining their behaviors and habitat choices. She once took a reporter out with her, and while discussing things later at a BBQ joint, he admitted, "I could not do this," simultaneously scratching his many bug bites. Jayme sometimes found herself "crawling under dense vegetation—it was incredibly silly," looking at snake-eye level for her subjects. She attributes her success to being "a natural hunter—a scouter I should say. Snakes are hard to find."

Her studies solved many mysteries about the eastern diamondback. "What I'm most proud of is we illustrated the importance of the pine savannah to this snake. It doesn't matter what scale you look at—it's a pine savannah specialist." She described how, before her study, many assumed the snakes hid underground during the summer months. Instead, she found them camouflaged in the savannah grasses, and they were simply impossible to find without telemetry. These massive rattlesnakes let you walk right past them, and you never know they're there.

Australia. But these are not the deadliest snakes. Nor are the seasnakes, which are admittedly very toxic.

To understand which snakes are the deadliest, we should first define our terms. The most *toxic* snakes on Earth are in fact the Australian elapids and their close relatives, the seasnakes. The toxicity of snake venom is usually determined by injecting it into mice, and counting how many die as a result. Australian snakes have potent venom, a tiny amount of which can kill a lot of mice. However, because of Australia's remoteness, its correspondingly low human population densities, and its state-of-the-art antivenom and education programs, few people are bitten in Australia in any given year, and most who do survive the bite. So, although it is possibly the most toxic snake on Earth, the inland taipan (also known as the "fierce snake") has no recorded fatalities that I'm aware of. This is because it is found in extremely remote areas of Australia, undergoes dramatic population fluctuations that track rodent numbers, and it's actually a rather inoffensive snake, tending to retreat if approached. Therefore, although toxic, the inland taipan is not the deadliest snake in the world.

There is another criterion for judging the deadliest snakes, and that is how *dangerous* they are. I define a dangerous snake as one with a large amount of potent venom that's not afraid to use it—a combination of toxicity, size, and what people consider "aggressiveness." Although a more accurate way to describe it would be "defensive," we can agree that such a snake should be fairly grumpy, with toxic venom, and have an efficient venom delivery system that would allow it to easily inject venom into a human. In other words, which snake would you least like to be locked in a closet with? Toxic snakes such as seasnakes and coralsnakes do not fit this concept. Their small fangs and (usually) docile nature make them pretty harmless. You're not going to be walking through a rainforest with rubber boots on, step on a coralsnake, and have it bite you through the boot on the foot. The black mamba and king cobra are the largest venomous snakes in Africa and Asia, respectively, and both have large gapes and fangs allowing efficient venom delivery. Both have copious amounts of extremely potent venom. Both co-occur with a lethal menagerie of furry and feathered predators, and so they are willing to defend themselves from enemies with awesome dexterity. I would not want to be locked in a closet with either one. Nor would I necessarily want to be in the same building with them. But neither are particularly common in areas with large human populations, and these are not the deadliest snakes on Earth.

What makes snakes deadly is instead a combination of biological attributes and human demographic and socioeconomic factors. The world's deadliest snakes are species most people have never heard of. The two *deadliest* snakes on Earth—the ones responsible for the most human fatalities—are the saw-scaled and Russell's viper. They trade turns one year to the next as the reigning champ. And the number of people killed by these snakes is not insignificant. In the United States, perhaps a dozen people a year are killed by venomous snakes. In parts of Africa and Asia, tens of thousands die each year from snakebites. But the reason for this high body count has as much or more to do with humans as with snakes. These vipers occur in regions with high population densities of poor people, and a huge factor involved in their lethality is the simple fact that a lot of these people walk around without shoes. The other crucial factor is lack of access to modern medical facilities. Both species are among the most toxic vipers, and both species dislike being tread upon. The saw-scaled viper is cryptic, lying coiled in the sand, and will bite with little provocation if stepped on it or near it. So biologists must reluctantly admit that, worldwide, snakes are a public health concern. And most envenomations and fatalities throughout the world are unprovoked—cases where some poor farmer is walking along without shoes, minding his own business, and is bitten on the foot. Owing to the poor medical care and antivenom programs in these regions, the bites are frequently fatal.

But this is positively not the case in the United States. Unprovoked snakebites are referred to in the medical literature as "legitimate," and bites that result from harassment of the snake are considered "illegitimate." Almost all of the snake bites in the United States are illegitimate and occur during the attempted dispatch of venomous snakes.

This wasn't always the case. The epidemiology of snakebite in the United States once fit the pattern typical of developing nations. Circumstances of 71 snakebite fatalities in the United States from 1950 to 1954 reveal that typical victims were children, including a large number under 4 years old. The next most vulnerable group was elderly folks. Most people were bitten close to home, especially near farms. Many of these people did not receive prompt treatment, which during that time was primitive. Over half the victims were bitten on the lower extremities.

These data contrast starkly with recent data, which show first of all that snakebite fatalities have decreased drastically since the 1950s. Snakes caused about 30% of deaths attributable to venomous animals from 1950 to 1959, and dropped to below 10% from 1979 to 2007. Now most people are bitten on the upper extremities. Medical professionals have developed a profile of the typical American snakebite victim: Southern, white males aged 15–25, and a large portion of bites inflict the hand. Consumption of alcohol is a significant risk factor. I hesitate to feel sorry for such people. Consider also a letter written by two Phoenix physicians to the *New England Journal of Medicine* warning the medical community of the dangers posed by *dead* rattlesnakes. They reported no fewer than 15% of the patients they treated for snakebite were bitten by a snake the victim thought was dead. One case is a particularly good illustration: "Patient 3 shot and then decapitated a rattlesnake. His right index finger was envenomated when he picked up the head. He developed a self-limited coagulopathy but ultimately required finger amputation." He should feel lucky. He only lost his finger, but the snake lost its head.

Which of America's snakes is the "deadliest?" By our established criterion, we could list one of two snakes as the most toxic. Both the Mojave rattlesnake and tiger rattlesnake of the American Southwest have potent venom, at least in terms of its ability to kill mice. The Mojave rattlesnake is quite a bit larger than the tiger rattler, however, and it has a much higher venom yield, making a bite from this species much more serious. Fortunately, Mojave rattlesnakes prefer living in remote desert locations, and bites from this species are rare. Coralsnakes have potent venom because they belong to the same family as cobras. But coralsnakes should not be considered dangerous because they are a small species with a small potential venom yield, an inefficient venom delivery apparatus, and they are uncommon. I have searched far and wide for coralsnakes in the United States and have never seen one. Only one fatality has been reported for an eastern coralsnake bite in the past 40 years. The "victim" was drunk and attempting to kill the snake with a broken bottle. The Arizona coralsnake has never been responsible for any reported fatalities.

Besides the rather large and toxic Mojave rattlesnake, our most dangerous snake is probably the eastern diamondback rattlesnake. Although it is now much rarer than it was in the past, it still lives in some areas with fairly large densities of people, and some of these people actively hunt this snake for rat-

Tiger rattlesnakes have potent venom that can kill many mice. *Photograph by Pierson Hill*

The Mojave rattlesnake is a large species with strange venom containing neurotoxic properties. *Photograph by Todd Pierson*

Coralsnakes have potent neurotoxic venom but an inefficient venom delivery system. They are usually docile and uncommon. Only one fatality has been attributed to a coralsnake in the United States in the past 40 years. *Photograph by Noah Fields*

tlesnake roundups or for religious purposes. It is the largest rattlesnake and has the largest venom yield of any venomous snake in North America, and it will stand its ground in self-defense if provoked. Timber rattlesnakes have a high venom yield, and their venom is actually more potent than that of their larger cousin. They also live in the densely populated East, but they too have been eliminated from much of their former distribution, and they are a docile species. Northern and southern Pacific rattlesnakes are a medium-sized species with potent venom but are likewise inoffensive. They live throughout the overpopulated California coast, even within the city limits of Los Angeles.

The eastern diamondback rattlesnake is our largest venomous snake, and therefore has the largest venom yield. It is dangerous and has been responsible for many fatalities. *Photograph by Daniel Wakefield*

Timber rattlesnakes are dangerous snakes found in some heavily populated areas. *Photograph by Pierson Hill*

The southern Pacific rattlesnake is fairly large and has potent venom. This individual was photographed in an upscale gated community in Malibu. *Photograph by Mark Herr*

The western diamondback rattlesnake. *Photograph by Dirk Stevenson*

Western diamondback rattlesnakes are another large species preferring habitat that was once home only to jackrabbits, mesquite, and creosote. Americans have managed to set up cities where this snake lives, and the large urban centers in the Southwest—Tucson, Phoenix, San Antonio, and Dallas—have populations of this species close by. The eastern and western diamondbacks as well as the timber and Pacific rattlers are the species most often responsible for fatalities in the United States.

But, again, the reason why has much to do with their proximity to large populations of humans, and if they were left alone, they wouldn't bite anyone. This is certainly the case with the species responsible for the most envenomations in the United States each year. The copperhead persists across the Eastern Seaboard in relatively high numbers despite the huge urban centers present there. I have found them within the city limits of Atlanta, and they occur in or near other major cities as far north as the outskirts of Boston. They are responsible for nearly as many bites each year as all the rattlesnake species combined—all of the rarest of which are nonlethal.

Snakes Don't Attack People

You may be asking yourself how this could be, after all the movies you've seen, after all the excellent "documentaries" you've watched on Animal Planet, and after all the stories you've heard about how dangerous venomous snakes are. A good way to get a true feel for how unthreatening venomous snakes are is to go see it for yourself. But this is something I'm not going to recommend. Instead, I'm going to describe it to you, as I've done with my many close calls that I mentioned earlier. Snakes are not lightning fast, and don't slither up to you, chase you, or attack you. It is a fact that pitvipers have a quick strike. But their strike is given from a position of immobility, with their fat bodies anchoring them to the ground. You can be a yard away from a 5-foot-long eastern diamondback and it can't get you. And their stout bodies make them slow, with the exception of the strike. They can't dash forward around a pole and climb two stories in milliseconds like you saw in the movie *Anaconda*. They don't attack people. They stand their ground and will strike from a sitting position if you mess with them. People attack snakes, and snakes defend themselves. Snakes do not cross the distance between themselves and a person to attack.

But there are legends aplenty about snakes that attack people. The cottonmouth is the American snake most frequently accused of this behavior, and king cobras, black mambas, and Australia's eastern brownsnakes (no relation to America's harmless Dekay's brownsnake) are also reputed to cross the distance between themselves and a human. I have even heard stories to this effect from reputable herpetologists. For my part, I can offer my observations

The copperhead is responsible for the most venomous bites of any American snake. Fortunately, its venom is mild, and most bites are not fatal. *Photograph by Curtis Callaway*

of nearly a thousand cottonmouths to assure you that they never once tried to come near me. They would vigorously defend themselves when I tried to catch them, and they would usually try to escape when I saw them. Sometimes they'd stand their ground and pop open their mouths in a threat display, but that's it. All the stories about cottonmouth nests, cottonmouths falling into boats, cottonmouths chasing people, and cottonmouths being aggressive are totally wrong.

Bruce Means published the only scientific paper about snake attacks, and he concluded that snakes perceived to be attacking people are in fact attempting to escape. He described a few incidences in two of the species most often described as attackers: Australia's eastern brownsnake and the American cottonmouth. He describes these snakes coming right at him as he was backing away. For both species, if he sidestepped them, they continued past him and did not continue to "chase" or bite him. In the case of the cottonmouth, he could provoke this behavior by stepping between a cottonmouth and their closest aquatic escape route. The snake knew what it had to do to escape, and it barged right at him to get to the water:

> With its neck spread slightly, the snake continued to advance as I backpedaled, making a few striking parries as it neared my feet. When I stepped sideways from the snake's trajectory (which was directly toward a swamp on the other side of the road), I was gratified that the snake maintained its original direction and did not turn to follow (or "chase") me. Its "aggressive" behavior obviously was a bluff to assist the snake in making its getaway into the safety of the swamp.

I can also offer a few observations of snake "attacks" that could be misconstrued by somebody unfamiliar with snake behavior. Once I was canoeing and a cottonmouth drifted right past me in the current, and it even popped open its mouth in a threat display. Surely a fisherman terrified of snakes would have flipped his boat in a panic if that had happened. But I just watched it go by, and that it did. I've seen one copperhead—out of dozens I've seen on roads at night—strike viciously multiple times to defend itself. It did so with such enthusiasm that it lifted itself off the pavement a few centimeters with each strike, and it launched itself toward me each time by a couple of inches. Surely a motorist who stepped out of his vehicle to examine such a snake would have considered this rare encounter as evidence of how aggressive copperheads are. Often while looking for snakes we come across them out in the open near their burrow entrance. The snake will come right toward you to get to the burrow, and it will crawl between your legs if it has to. A large Australian copperhead (no relation to the American copperhead) approached me precisely in this way. It was not being aggressive. It was only trying to get away.

Unfortunately, most people in this country who get bit are asking for it. They are putting themselves in harm's way by intentionally trying to kill a venomous snake. But you wouldn't try to get out of your car to kill a raccoon, would you? I suspect that venomous snakes are about as dangerous to humans as raccoons are, but the perceived threat of each is incredibly disproportionate. Raccoons are carriers of rabies, a disease that is 100% fatal if not treated promptly. Rabid raccoons occasionally attack people unprovoked. Although only a small percentage of raccoons carry the disease, thousands of Americans would be exposed to rabies if people went out of their way to try to kill raccoons. Especially if they went after raccoons with garden implements, rakes, and axes, and then tried to cut trophies off the raccoons. If people treated raccoons with the same irrational disrespect as snakes, there would probably be dozens of deaths attributable to rabies each year, and there would be thousands more hospital visits for rabies treatment. Instead, we have thousands of visits to the emergency room for snakebite treatment, and only a dozen or so fatalities caused by snakes. Yet I suspect that most people who see raccoons in their yards, or crossing the road in front of their cars, view them with something ranging from healthy respect and curiosity to actual enjoyment. If you want to be safe from snakes, you simply need to have the same attitude toward them as you do all other wildlife.

So You've Been Bitten

Snake venom is a variable cocktail of proteins, enzymes, and polypeptides sculpted by natural selection to kill prey rapidly and begin the process of digestion. Think of venom as potent digestive juice. The actions of the venom depend on the kind of snake. Pitvipers usually have a large amount of hemolytic venom, which breaks down tissue and blood. Elapids usually have neurotoxic venom, which interferes with the ability of nerves to control muscle, but elapid bites are so rare in our country that this is hardly worth mentioning. And some rattlesnakes have a discrete and sizeable portion of neurotoxic venom along with the less dangerous hemolytic venom. The relative proportions and potency of each kind of venom lead to a great variety of possible outcomes of snake bites. You might receive a copperhead bite, which would dose you with small amounts of rather weak hemolytic venom. You'd suffer some pain and swelling, but you'd otherwise be fine. If you are bitten by a Mojave rattlesnake, you might not have much pain and swelling, but the neurotoxic venom could possibly send you into respiratory failure within an hour, the neurons controlling your lungs paralyzed.

As we have seen in chapter 2, pitvipers have an extremely effective venom injection system composed of hollow fangs that act like hypodermic needles. Coralsnakes have small fangs that are in a fixed position in the front of their mouth. The delivery system of pitvipers is the most effective of any snake in the world. It is so efficient that it can be metered—the snake can control how much it injects—and there is a possibility the snake will try to warn you first by giving a sporting defensive bite. If you're lucky, you may receive what is referred to as a "dry" bite. This will usually result in some discomfort but no serious symptoms. Something like 25% of pitviper bites are dry, odds that are a little better than Russian roulette. With coralsnakes the odds are better, probably because they must actually get to chewing on you to inject the venom, owing to their relatively inefficient injection system. But don't risk it. If you are bitten, go immediately to the hospital.

At the hospital, if you have received a more serious bite (probably because you riled up the snake), your symptoms will multiply. The swelling will become incredible. The enzymes will start breaking down your capillaries, causing fluid to build up between the outer and inner layers of your skin near the bite. Your own immune system will contribute to the swelling, as inflammatory factors and white blood cells struggle in vain to deactivate the venom. Tingling sensations—pins and needles—swarm your face, and you might have a strange metallic taste in your mouth. Waves of chills and nausea begin within the first hour, which will surely induce panic. If you receive antivenom promptly, this may be as bad as it gets—painful swelling and some alarming discoloration and blisters. But if you are slow to the hospital or if the venom travels quickly away from the site of the bite, you may face more general, so-called systemic, symptoms. Local symptoms are the effects centered near the bite. Systemic symptoms are those that affect your whole body. Heart rate and blood pressure abnormalities occur. Your blood begins to spontaneously clot within your veins, and capillaries accumulate clotting factors. When these reactions start to interfere with kidney function, kidney failure isn't far away. When it gets this bad, death often follows.

None of these symptoms happen within seconds, or even minutes. It's not like in the movies, where the cowboy is bitten and collapses dead within seconds. Untreated snakebites require several hours or even days to cause death. You have plenty of time to get to the hospital, and late treatment is better than no treatment and can still save your life.

Death happens only in the rarest of cases. More typical is the maiming that occurs due to tissue necrosis. Necrosis is tissue death—the venom digests the tissue beyond the point of healing, and the area around the bite becomes functionless and gooey. This frequently results in amputation. For the twenty-first-century American victim, this is most likely going to mean a lost finger and ugly scaring. I know some herpetologists and several snake enthusiasts who are missing parts of fingers. But this isn't the only possible legacy of snakebite. Some herpetologists I know have chronic pains and strange symptoms from their snakebites, some of which are decades old. Some have fingers that still tingle or never again work in quite the same way. One colleague developed a bizarre allergy to snakes after receiving a bite from a rattler, and can no longer work close to rattlesnakes without tearing up and sneezing.

Snake Cowboys

Most people are scared of snakes, and I'm fine with that. I'd prefer they were so scared they never went near them. That's what you need to do to stay safe from snakes. Don't go near them. This doesn't mean you need to stay indoors; my close calls with venomous snakes confirm that even if you are outdoorsy, you will be just fine.

Instead, an alarming new trend is the rise in snakebites occurring in people who love snakes. These are people who keep venomous snakes as pets, and people who go looking for them for the rush of catching them. These "snake cowboys" are nothing more than adrenaline junkies. They go out looking for "hot" snakes—typically only satisfied by the most dangerous kinds—and use

These snake enthusiasts are admiring a rattlesnake responsibly, photographing it from a safe distance. *Photograph by Mark Herr*

hooks to pick them up and "tail them." Their web pages are full of pictures of them looking manly, proudly holding each dangerous snake they found. These people also go to "hot snake" shows, and buy such ludicrous species as gaboon vipers and king cobras to keep in their homes. Some of these same people feel the need to "free handle" their snakes—hold them with their bare hands!—to prove their salt. These people do indeed love snakes, but for the wrong reasons.

The thing is, whether you love them or hate them, you need to give snakes respect. If more people respected snakes, the number of snakebites and snake-bite fatalities in this country would plummet to practically nothing. There really is no such thing as a dangerous snake. But boy are there some stupid people.

Snake-Proofing Your Yard

One of the most frequently asked questions I get is how to make a yard snake-proof. I usually remind people of the negligible risks posed by snakes, how snakes are part of the environment and should be left alone, and then admit with a forced, apologetic tone that there is nothing you can do. Snakes are too good at infiltrating tight spaces for snake-proofing to be possible. And they're vertebrates, so any chemical you try to put out as a repellent would also repel you, your kids, and your dogs and cats, too. There are some people who say that a layer of mothballs around your house will keep snakes away, but there is evidence that this method doesn't work. And eventually all that naphthalene is going to wash away and evaporate. For some folks, this just isn't good enough. They often seem offended that I can't give them better options, citing their children's safety. They cannot be convinced to live and let live when it comes to snakes.

There are some handy tricks you can use to reduce the appeal of your yard to snakes, but these are only half measures. Snakes do like to take cover under things, so you should remove brush piles, flat rocks, wood piles, your El Camino up on blocks, any old farming equipment, or hide objects that may be lying around your yard. In the old days, folks kept neatly manicured lawns that they burned once a year to keep snakes away. Mowing regularly is probably is good enough. And be sure the entire yard is just one tacky, monotonous lawn, with no shrubs or exposed areas that look remotely natural. Many snakes don't like open agricultural landscapes and actively avoid them. Another option would be to pave your yard over entirely, because certain tiny, harmless snakes may actually be *in* your lawn. Also, be sure to trim back any tree branches that are hanging over from adjacent properties. Tree-climbing snakes—most of which are harmless, by the way, but you asked for this—can easily crawl along thin branches and invade your property from your neighbor's property. But having an all-concrete lawn with no piles of wood and no aerial pathways or other possible hiding places will not guarantee a snake-proof yard. Some neighbor might have a log pile, trees, brush, and nice warm rocks in their yard. A harmless snake might be over there, and decide to crawl across your perfectly manicured concrete lawn.

What then? Well, I recommend installing a drift fence as a barrier to trespassing snakes. Drift fences work well for capturing snakes, and you can turn this to your advantage: snakes will crawl up to the fence and crawl along its edge, and so long as there is no gap, they won't be able to get in. Make sure it's made of aluminum flashing about 5 feet tall just in case a harmless indigo snake lives nearby, and sink it at least a foot into the ground in case a harmless, subterranean wormsnake lives in the vicinity. If you have a driveway, I'm going to recommend you install a sliding gate that opens and closes when you pull in, and be sure to be vigilant while parking—don't let in any snakes when you come home from work.

Because if you do, they will then be trapped inside your yard! Which brings me to the final, crucial phase of snake-proofing: you need to be sure that, before you install your drift fence snake barrier, there are no snakes *already on your property*. Otherwise you'll be stuck with them. In order to remove these freeloaders, you'll need to hire an expert. There are herpetologists in most major cities with zoos or universities, and I can get you in touch with them. Their rates vary, but I'd say you could probably hire a decent one for around $50,000. All told, the entire snake-proofing operation will look god-awful and cost you about $100,000. I know this sounds extreme and exorbitant. But this is the safety of your children we're talking about.

Snakebite Treatment and First Aid

There is no known reliable first-aid treatment of pitviper bites. Anything you heard about tourniquets, venom extractors, tasers, or ice is not based on modern medical practice, and some of these things can do more harm than good. Tourniquets stop blood flow to the extremities through the veins and arteries, but venom does not typically travel through your body through these routes. It travels directly through connective tissue and the lymphatic system, so it will get through anyway, and meanwhile you could cause real harm to the extremity you're trying to save. Once a man went to the hospital for snakebite, and when he got there, his hand was purple and lifeless. Doctors had to amputate it. The man brought in the snake that bit him for identification, and it turned out to be an eastern kingsnake. Eastern kingsnakes are a non-venomous species. He had tied a tourniquet as first aid for a harmless snakebite and it cost him his hand.

In Australia, there are no pitvipers, so they use pressure bandages for first aid. Pressure bandages are wrapped tightly to apply the pressure, but not tightly enough to cut off blood circulation. Australian venomous snakes are all elapids, and their venoms travel rather slowly through the network of lymph fluid and lymph nodes. There is evidence that pressure applied to the bitten extremity slows the circulation of the venom. In any case, this is not an effective treatment for pitviper bites, because pitviper venom rapidly breaks down tissues and will continue unimpeded whether you apply pressure or not.

Although there is little evidence of their effectiveness, venom extractors are coming back into vogue after a long hiatus. A recent news report described a man in Alabama whose "life was saved" because he used a venom extractor after being bitten by a timber rattler. The news story claimed a doctor attributed his survival to the extractor. Now, I don't believe local news stories any more than I believe most used car salesmen, but some recent medical literature says that using a venom extractor within 5-7 minutes of a bite may succeed in getting some of that venom out of you. Experts disagree, asserting that not only are the extractors ineffective, they could actually exacerbate the venom's effect. The title of a recent appraisal of the efficacy of venom extractors is revealing: "Snakebite Suction Devices Don't Remove Venom: They Just Suck." But you may have nothing to lose if you're 15 hours from a hospital (not likely anywhere in the United States) and you don't mind slapping down $10 for a dubious piece of plastic equipment ahead of time.

The best treatment for snakebite is found at the hospital, so be sure to get a ride to the hospital if you're bitten. Snake experts have been saying for years the best first aid for snakebite is a set of car keys, and it's still true. Even better, if you have a phone, call 911, and the dispatcher will decide whether to send an ambulance and will coordinate the treatment. You've got plenty of time, so don't panic. You can try elevating the bitten extremity, which will help control circulation a little. When you get to the hospital, the doctor will decide how dangerous the bite is, and by the time you get there it should be pretty obvious. You don't even need to know what kind of snake bit you, because doctors use the same treatment (a product known as Crofab) for all pitviper bites. They will start giving you antivenom, which is a bunch of reconstituted antibodies that are grown in a sheep. These antibodies will wrap up the venom proteins and deactivate them. Additional treatment for various symptoms may be required. For a copperhead bite they may not even administer antivenom, because it's expensive and antivenom can cause an allergic reaction in some patients. You better hope you've got insurance, because all this stuff is going to cost you. People have been slapped with bills costing tens of thousands of dollars after snakebite treatment, so be sure you have good insurance and a low deductible if you plan on harassing a snake. They're not going to turn you away if you don't have insurance, but after receiving the bill you may feel like dying. Yet another reason to avoid molesting snakes.

10 · Snake Invaders

The Ever Glades are now suitable only for the haunt of noxious vermin, or the resort of pestilential reptiles.

BUCKINGHAM SMITH, 1848, from the first report to the US Congress about the Everglades

And there was a war in heaven. Michael and his angels fought against the big snake. And the big snake and his angels fought back.

BOOK OF REVELATION

Pestilential Reptiles

In 2003, Everglades National Park ranger Gary Landry was driving down Lower Wagon Wheel Road on his ordinary rounds. For law enforcement rangers in the Everglades, this beat typically included ticketing flagrant speeders ignoring the 50 mph speed limit that protects marsh rabbits, white-tailed deer, and the last Florida panthers. Being so close to Miami, rangers also dealt with diverse symptoms of modern society. Gary told me that when he worked in the glades, "there wasn't much going on," but that he "heard stories about back in the 80s . . . that was the time to be there," when pallets of narcotics dropped from low-flying planes, and rangers found burning cars with bodies. When Gary was there, he mainly chased poachers, where "the biggest thing was patrolling thousands of acres in airboats, when the locals know the routes like the back of their hands." They chased poachers who used airboats as battering rams to kill deer. They'd respond to alerts of "shots fired" only to discover people who just wanted to go out in the middle of nowhere to shoot guns in the air. Occasionally he'd find a small bundle hanging in a tree: a white sheet wrapped in a red one containing a dead chicken and spices. A Santeria offering. But nothing in his experience could have prepared Landry for what he would find that morning in 2003.

Lower Wagon Wheel is too narrow for cars and is used only by bicyclers and a tourist tram. On either side of the road, sawgrass extends to an infinite horizon, broken only by circular cypress domes. Off the shoulder, there are usually great egrets posed like lawn ornaments facing the shallow water. Occasionally there are rarer sights: ugly and magnificent wood storks perched like gremlins, alligators stretched out basking, and perhaps a raccoon shuffling along.

Gary received a call from the tram driver that there was a giant snake on the road up ahead, and suggested he get down there to catch it. Gary Landry is typical of many of our nation's law enforcement: brave, professional, and dedicated. But Gary is not a snake guy. Landry isn't scared of snakes, but his previous experience with them was catching harmless gartersnakes and seeing peo-

Gary Landry and friend in 2003. *Photograph by Bob DeGross, courtesy of the US Park Service*

ple with pet snakes around their necks; at the time, he said, "I was pretty naive about pythons. I always thought they were cool—they are so docile." When he got to the spot, the snake was coiled in the grass off the road. It was the biggest snake he'd ever seen. He walked out and grabbed the 9-foot snake by the tail. It reached around and lunged to bite his boot. Wild Burmese pythons are no joke; although nonvenomous, they are extremely dangerous and can inflict gruesome bites. Quickly sobered, Landry tried to get the snake's head under control with a pair of tongs, telling me later, "It's surprising how strong they are." Another ranger snapped a picture of Landry with the prize before he got it in a bag.

The photo shows Gary with a slightly goofy smile that seems to say, "what do I do with this thing now?" That photograph found its way into newspaper articles and scientific presentations all over the world. It was just one of many sensational encounters with Burmese pythons at the height of their invasion of south Florida.

Phase 1: Establishment and Lag Phase

Burmese pythons are native to Southeast Asia from Pakistan to southern China and Indonesia. They are among the world's largest snakes, with a maximum size of 27 feet. Only reticulated pythons and green anacondas get bigger. In their native range they feed on birds and mammals, ranging in size from rats, peacocks, pigeons, and ducks to porcupines, monkeys, wild pigs, deer, and antelope. They have not been confirmed to have killed and consumed humans in the wild, but at adult size they are capable of doing so. In the 1970s, Burmese pythons began turning up in the pet trade. The selective breeding of a mutant yellowish-color morph increased their popularity. Burmese pythons, or simply "Burms," became one of the most popular large constrictors in the pet trade. Sold as manageably sized hatchlings and juveniles, these snakes

Portrait of a south Florida Burmese python.
Photograph by J. D. Willson

quickly grew to 10 feet long and longer. Because they are long lived, many hobbyists eventually became bored and burdened with their pythons, which require equally large enclosures and a steady diet of rabbits and other expensive food. Burmese pythons can also have a nasty disposition; one day they are tame, trustfully hefted onto the shoulder of their keepers, and the next day they are lunging out menacingly from their cage. A few have constricted their owners to death. When I was a keeper at Zoo Atlanta, I received several calls from python owners who could no longer care for their snakes. I had to turn them away; we didn't like to display "pet store snakes," and we did not have the room for two dozen 16-foot pythons. Our suggestion was to euthanize the snakes, because the other option available to irresponsible pet owners led to an ecological catastrophe in south Florida.

The domestic flight from Dallas to Miami is a hall-of-fame flyover of American environmental disasters. While gaining cruising altitude, I had intimate views of the Trinity River Project, Dallas's attempt to rebrand one of the most polluted rivers in the United States. The high-water marks of the summer 2015 floods—barely contained within the city's levees—were like coffee stains on local farmland. Turning southeast, we crossed the Red River and followed it on to the Mississippi River Control Structure, a tenuous attempt to prevent intimacy between the Mississippi and Atchafalaya. This and other "improvements" of the Mississippi, including distributary shutoffs and its virtual entombment within levees, has only increased the likelihood of capture of the main stem Mississippi by the Atchafalaya, which will result in a new mouth a hundred miles to the west, and certain death for Baton Rouge and New Orleans. We continued south over toxic Lake Pontchartrain and the Superdome, which no longer bring to mind the New Orleans Saints and Mardi Gras, but instead images of floating bodies and distraught folks awaiting help. We followed the Mississippi River Gulf Outlet Canal into the Gulf. This giant, expensive dredging project, little used for shipping, provided Hurricane Katrina's storm surge a direct route into the city. Called the "hurricane highway," it's now closed off and armed with storm walls.

We passed along the edge of the great claw of the Mississippi Delta, which, thanks to the levees, is pumping billions of tons of brown silt far out into the Gulf instead of distributing it slowly for wetland formation. The Gulf glittered with oil platforms, reminding me of that big hole in the bottom of the Gulf disgorging crude all of summer 2010. I thought of the tenfold increase in the number of beached Kemp's ridley sea turtles immediately following the Deepwater Horizon disaster.

I dozed off and awoke in time to see the Big Cypress pass under the plane: a universe of shallow water pocked with perfect rings of a thousand cypress domes. The plane descended. We passed over the Shark River Slough: flat water reflecting blue gray in the slanted January twilight, buoying a flotilla of evergreen tree islands shaped like willow leaves oriented southwest. Big Cypress stretched on below us for miles in every direction, as the plane continued on for some 30 minutes. We then we flew over a level-straight canal, and below it the slough became disorganized; the tree islands dissolved into a mess. The whole system was once a "river of grass" flowing slowly across the width of south Florida from Lake Okeechobee to the sea. Then every attempt was made to strangle the natural flow, drain it off, and harness the dubious agricultural potential of the Everglades. Now we're attempting to reestablish the flow. At $8.2 billion, the Comprehensive Everglades Restoration Plan is the most expensive environmental restoration project in history.

We then passed the abrupt manmade frontier of the Everglades, over trenches and bulwarks, canals and wastewater treatment plants, rows of Australian pine and melalueca, subdivisions and interstates, and I closed my eyes as the jet touched down at Miami International.

I traveled to south Florida in January 2016 to take part in the Python Challenge, an annual python-catching jamboree hosted reluctantly by the Florida Fish and Wildlife Commission. Enthusiastic snake catchers ranging from lawyers and suburban dads to PhD herpetologists form teams and scour the swamps, marshes, and roads of south Florida looking for the invasive snakes. Wildlife officials are aware that an annual hunt will do little to control the pythons, and instead view the event as a public relations opportunity. But early returns from the first events were worrying: participants typically returned home after a day or two without seeing any snakes, convinced that there was no longer a python problem. In 2016, event organizers were certain to point out that, despite their size and ecological impacts, the pythons are extremely hard to find. The 2013 python challenge included 1,600 participants who found only 68 snakes. This year promised to be the biggest python challenge yet, and my friends and I were going to be in on it. Before arriving in south Florida, we exchanged e-mails predicting how many pythons we'd find. The estimate ranged from 12 to 24, and as far as the possibility that we might completely strike out, my friend Sean Sterrett simply responded, "that's not an option."

I flew down early to meet up with Skip Snow, a Park Service biologist who witnessed the python invasion firsthand. We met up in Everglades National Park and had a conversation along a boardwalk trail busy with school groups.

Skip is in his sixties, tall and fit with a white goatee and white hair framing a tan, narrow face. His legs look like he's kept up with running all his life, but he explained that, now that he's retired, he mostly just walks and occasionally kayaks Biscayne Bay. He arrived in the Everglades in 1988 and worked on various projects until 2000, when the park began efforts to document and manage its many introduced species. Soon after he began surveying and coordinating research on exotics, Burmese pythons began their dramatic population boom.

Unbeknownst to Skip, pythons were already well established long before he first became involved. They may have been there as early as the time he arrived in south Florida. On this point opinions differ wildly, and we may never know exactly when and how the first pythons arrived in the Everglades. Population growth models and capture records support the view that a small founder population was established in the 1980s by irresponsible pet owners who released snakes near Flamingo in the southern Everglades. The pet lobby supports the claim that pythons escaped from a breeding facility near the Everglades when it was destroyed by hurricane Andrew in 1992. Deferring to the "hand of God," the pet lobby denies responsibility, and continues to downplay the effects of pythons. When efforts to ban other large constrictors from importation and trade were initiated, the pet lobby mobilized, using the standard playbook of industries fearing regulation: they warned of loss of revenue and profits, bemoaned impending job losses, and appealed to the right of all Americans to own whichever giant snakes they damn well please.

Regardless of the release site, the reason why Burmese pythons are in Florida is because the United States has a permissive and lucrative pet trade. Over 300 million individual frogs, lizards, turtles, snakes, birds, and mammals belonging to some 4,200 species were imported legally into the United States between 1968 and 2006. Nearly half of these were wild caught, which severely affects native populations. We export another 13 million each year all over the world. This does not include the annual $800 million global trade in aquarium species involving some 400 million fish. And we can only guess the impact of

Burmese python female brooding her eggs. *Photograph by Mike Rochford*

the black-market trade in animals. Given that about 10% of species introductions lead to established populations, it is remarkable that we are not overrun with more exotics.

Most invasive species undergo a lag period when they first become established: population sizes remain low for decades as the population becomes adjusted to its new habitat and individuals fumble around trying to find one another. Population growth models suggest that during this initial phase, Burmese pythons were present only locally near their initial release site. Walter Meshaka, then curator of the Everglades National Park natural history museum, began finding pythons in the 1980s. Throughout the 1990s, only small numbers—perhaps only two or three a year—were found in the park, mostly on the main park road and near its southernmost tip among the mangrove swamps near Flamingo. Based on these observations, Meshaka made the bold claim that Burmese pythons were established in the Everglades, and that a small breeding population was present and growing. Almost everyone scoffed at this idea, instead brushing off the small numbers of pythons as recently released pets doomed to die of starvation. Soon the occasional south Florida winter frosts would kill them off. Or fire ants would wipe them out. Or alligators would eat them all up. So continual python sightings were not seen as a problem and were instead likened to alligators showing up in New York City sewers.

Phase 2: Exponential Growth

In 2002, the *Miami Herald* reported a large python found by a fisherman in the Everglades backcountry some 40 miles up the Shark River Slough. This was far from Flamingo and the main park road, and the story caused Skip Snow's expressive eyebrows to raise high on his forehead. This finding did not fit the pattern of small numbers of pet pythons being released by park visitors. According to ranger Raymond Little, between 2000 and 2003, hunters in Big Cypress began finding pythons "way back in the backcountry," and soon "it got to the point when we saw pythons almost daily."

In January 2003—the same month Gary Landry found the giant python on Lower Wagon Wheel Road—Skip got a call from rangers about a large alligator engaged in a pitched battle with a giant snake. He headed to the Royal Palm trailhead and raced across the boardwalk to a large, dark pool among tropical hardwood trees and cypresses. We now stood in the same spot as he recalled the drama: The snake was wrapped around the alligator, and the alligator had most of the snake's tail in its mouth. They trashed around violently for hours. After some time, he thought the snake was dead, but when the alligator loosened its grip, the python "came to life like the Phoenix" and the battle resumed. The park superintendent—Skip's boss—arrived with his entire family and they witnessed the battle firsthand. There was no longer any doubt that something strange was happening in the Everglades.

Reporters arrived and took pictures and video. Everyone had plenty of time to see the battle because it went on for 30 hours. Images of the alligator wrapped in a python went viral, circling the world on the Internet. Skip's phone rang every 10 minutes with news reporters requesting interviews. The superintendent called to ask if he needed any additional park resources. The next day, the alligator finally released its grip on the python, which Skip now thought was finally dead. He assumed the alligator would finally eat it. Instead, the python miraculously revived and slithered away, the battle ending in a stalemate.

Later in 2003, numerous baby pythons were found in several locations within the park, including out behind Skip's office.

In 2004, several adult pythons were killed by mowers maintaining water retention berms just outside the park.

In 2005, another gator battle made an even bigger sensation and spawned a *National Geographic* special: a photographer was flying over the Everglades in a helicopter, happened to spot a strange object, and angled down for a closer look. He landed the amphibious craft and found a large, dead alligator exploding from the stomach of a dead python. The python must have eaten an alligator that was just too big and died, and the dead alligator eventually emerged from the decomposing snake. When Skip first began telling me of the 2003 gator battle, I assumed he was speaking of this one. Instead he turned away, shaking his head, and told me "No. Not the exploding python battle. There's so many. So many."

In 2006, the first nesting python was found under a carpet of roots by a professional snake catcher. Burmese pythons have an adaptation that gives them an edge in cooler climates: females actually attend their clutch of eggs and rapidly contract their body muscles, raising their body temperature, warming the clutch, and speeding development.

The pythons were entering the next phase of invasion. Long ago, Darwin borrowed the concept of exponential growth from the economist Thomas Malthus. For Darwin, exponential growth provided the huge numbers of superfluous individuals within a population that are culled by the unflinching hand of natural selection. It is unchecked population growth. Applied to python populations, exponential growth describes the huge numbers of snakes you can expect if every single egg hatches and lives long enough to breed. Once established, many invasive species experience this horrifying expansion because they have no natural enemies. Instead, pythons found south Florida to be an equitable climate full of prey of their preferred size. They began growing quickly and feeding frequently. They put on weight fast and began to mature sexually. Then they began laying eggs. Pythons can lay 18–85 large eggs, with an average of 40. This is a reproductive capacity far greater than most of our native snakes. And, lacking effective predators, diseases, or competitors, the eggs hatched into 2-foot-long babies (larger than any native newborn snake), which began feeding on small rodents and other prey. Many of these then lived long enough to reproduce.

We can do some simple math. Let's say a single female lays 40 eggs. Of these, given an even sex ratio, about 20 are female. If all 20 of this first generation find a mate and lay 40 eggs, that quickly results in 800 additional pythons, of which 400 will be female. If these lay 40 eggs apiece, and all of these become adults, we're talking about 16,000 pythons in just two generations. The third

A Burmese python crawls along a wetland in the Everglades. *Photograph by Noah Fields*

generation gives you 320,000 pythons. The fourth gives you 6 million. That's how exponential growth works. And that's how pythons went from a doubtable problem in the 1990s to an environmental catastrophe a decade later.

Phase 3: Carrying Capacity

Skip threw everything at his attempts to study and capture pythons. In 2003, he initiated a 61-point action plan establishing public service announcements, a python hotline, regular surveys, and numerous other studies. Folks sent e-mails and letters suggesting ways to find and kill snakes. His favorite was from a Floridian who suggested building platforms out in the glades. The platforms were to be provisioned with a goat and a pig. The goat served to attract the pythons, like the Tyrannosaurus from the movie *Jurassic Park*. But, unlike the dinosaurs, which have no natural enemies, the snakes would be quickly dispatched by the pigs, which every Floridian knows are effective killers of snakes. He got offers for help from the US Armed Forces. He met with an advisor for US Special Operations Command, who brought state-of-the-art infrared and night-vision goggles. If it worked, the strategy would then be to "call in the troops and shoot them from the air." Skip remembers cruising down the main park road with the headlights out, scanning the sawgrass for thermal signatures through Special Forces optics. The best they could do was spot a bird perched in a tree, whose opaque outline appeared "like the moon from behind clouds." A navy man offered his help, scoffed at the army's equipment, and assured Skip that the navy had the best toys. Nothing worked.

Skip sought the advice of "card-carrying herpetologists," contacting Gordon Rodda and Bob Reed, two biologists who were already hip-deep in invasive snakes. They headed the effort to study and control the invasive brown treesnake, an Australasian species accidentally introduced to the US island territory of Guam as a stowaway back in the 1950s. The impact of this snake in Guam foreshadows the eventuality for pythons in south Florida. The island once had a small but interesting bird fauna found nowhere else in the world. The snake changed all that. Native birds, lizards, and bat populations have crashed since the brown treesnake invasion, and several species are being pushed to extinction.

With suggestions and help from Rodda, Reed, and many other snake experts, Skip coordinated a research program geared toward determining how prevalent the pythons were; what impacts they were having; and what, if anything, could be done to control them. They caught snakes and implanted them with radio transmitters to study their behavior and movements. There was still hope the pythons were only found in human landscapes within the Everglades, like canals and road levees. They tried using traps to catch pythons. Reed and Rodda used climate data from the python's native range to project how far the pythons might spread in the United States. None of the results were encouraging. Telemetry studies showed the snakes were adjusting to

A brown treesnake, an insidious, invasive species introduced to the US territory of Guam. *Photograph by Crystal Kelehear*

A Brahminy blindsnake, an exotic species found in Florida, Georgia, and Hawaii, whose effect on native ecosystems is unknown. *Photograph by Pierson Hill*

their new habitat with relish. Pythons move huge distances and were tracked far away from roads to remote corners of the Everglades. The cold winters of south Florida—including the deep freeze of 2010—did little to slow the invasion; some pythons died, but many survived. They simply sunned themselves during the colder months and could flatten sawgrass into a purpose-built basking platform. The traps were ineffective, although they did manage to catch a few ratsnakes. And Reed and Rodda's climate model stunned the world and created a sizeable controversy, because it projected that pythons might do quite well in cooler climates to the north. In fact, it predicted the expansion of Burmese pythons into the southern United States as far north as the Delmarva Peninsula.

The most discouraging news came from diet studies. Counts of mammals along park roads decreased dramatically during the python explosion. Opossums and raccoons, once common in the Everglades, are now rarely seen by park visitors. Marsh rabbits have been practically eradicated. Nearly every native mammal and bird in the area has been found within the gut of a python: round-tailed muskrats, squirrels, rodents, rails, limpkins, foxes, raccoons, bobcats, ibises, herons, egrets, wood storks, magnificent frigatebirds, and white-tailed deer. One python was found with 14 cotton rats in its gut. The Key Largo woodrat, an endangered species found nowhere else in the world, has fallen prey to pythons, and concern is mounting that pythons will drive them to extinction. It is only a matter of time before one of Florida's last panthers finds its way into the belly of a python.

Our python-catching team consisted of me and my friends Dave Steen, Sean Sterrett, and Stephen Neslage. We arrived on the first day of the event and waited in line to enter one of south Florida's water conservation areas: a mosaic of sawgrass flats, hardwood hammocks, and canals that we knew was teaming with pythons. While we waited, I talked with some of the other participants to gauge our competition. Besides spending time together for a guys' weekend, the main motivation for participation in the challenge appeared to be father-son bonding. Two such pairs were waiting in line with us, including a duo from Orlando who had been waiting patiently since 5:00 a.m. They were both well versed in python biology from heavy consumption of cable nature show documentaries, and selected this hunting spot after consulting a website showing all python captures. I returned to my team and could barely contain myself, giggling that these two rubes were planning on catching pythons with a dog catcher's noose. The other father-son team was armed with a .306 equipped with telescopic sights. Apparently unaware of the contest rules prohibiting firearms, their plan was to shoot the pythons from across the canals. This was shaping up to be quite an event.

We slowly and deliberately paced the canals for two days. The canals support a gallery forest of subtropical hardwoods and occasional whisk ferns—primitive plants somewhere between real ferns and mosses. Sean and I even waded out to a few nearby hardwood hammocks in the hopes of boosting our chances. We waded through thigh-deep water among sawgrass and lumpy clouds of organic ooze and found a cottonmouth and a circular bed of matted-down sawgrass at the edge of a hammock. If we had been anywhere else, I would have assumed it was the bed of a deer. But this is python country now.

At the end of the first day, we bumped into the dog catchers, who were beaming with pride. They had captured a python. We listened with thinly veiled fury as the bumbling, lisping suburbanite explained how he looked down into the thick grass on a canal and suddenly noticed a 6-foot python between his legs. He backed away and started texting his son to come over for the assist. He couldn't get any messages through, so he bravely took it upon himself to catch the snake. With a dog catcher's noose. Proof of the noose's efficacy weighed down the cloth bag in his hand. We each secretly hoped he had somehow caught and misidentified a ratsnake. But to our great dismay, father and son opened the bag at the check station and proudly revealed a calm, beautiful Burmese python. Later we calculated the collective snake experience of our

Participants in the 2016 python challenge turn in their python to Florida Fish and Game officials. *Photograph by the author*

team relative to Team Dog Catcher. We had something like 36 years of experience catching and studying snakes, and they had none. There is no substitute for beginner's luck.

After another day of searching, it was almost time to return to Texas to start the new semester. I had one more afternoon before heading back to the airport. I also had to find a shower at some point so I wouldn't end up on a long domestic flight smelling like an ape. So I searched once more in earnest, gave all my friends a big hug, and walked to my rental car.

No sooner had I turned the ignition did I receive a text message indicating that Sean found a python. It was an 8-footer and gave him a nasty bite. It was an epic capture and he was all alone when he found it. Sometimes that's just how snake hunting goes.

I drove to the YMCA in Homestead and showered up. As I walked back to my car, a flash of movement caught my eye: an 8-inch lizard shot across the gravel sidewalk and up the trunk of a palm tree. I slowly scanned the side of the building and counted a few dozen lizards of the genus *Agama*, which is native to Africa. This population is currently localized in Homestead and has

Sean Sterrett with his triumphant python. *Photograph by David Steen*

not yet reached the exponential growth phase. There are dozens of such exotic amphibian and reptile species restricted to their point of release in south Florida, which may someday initiate range expansion. Some are large constrictors that are staying put for now: boa constrictors and African rock pythons are both present and breeding in a few spots in south Florida. Dozens more escape or are released every year; Miami-Dade County alone is home to 41 species of exotic amphibians and reptiles. That's not including formidable species that have escaped but have not yet become established, such as green anacondas, reticulated pythons, and king cobras. Meanwhile, the Burmese python continues to expand its range, and nobody is quite sure how far they will spread and when their populations will stabilize.

I made my way back up I-95 to Miami and stared out over the golf course lagoons, drainage canals, and residential lawns, imagining all the pythons, iguanas, anoles, geckos, and tegus living out there. If I had some sort of detector that placed a fluorescent tracer to indicate each one, the whole place would be lit up like neon. It's all python country now.

After working on the python problem for over a decade, Skip isn't optimistic about controlling their populations or preventing their expansion. South Florida is too wild and inaccessible. Exponential growth is too difficult to control. The pythons are too good at hiding. I suggested that perhaps if python skins were worth $1,000 maybe we could wipe them out. Aren't we good at wiping out species? Didn't we manage to wipe out the gray wolf and Florida panther? He had heard such suggestions before and politely shrugged off my naiveté. Instead, he thinks the best we can hope is for the Burmese python to become a "spokessnake" for the greater problem: the uncontrolled spread of invasive species. If anything good comes from the invasion, it will be that people finally realize the extent of the problem. It is past time that the United States began instituting stringent biosecurity.

Invasive species and their companion diseases are a massive problem. The list goes on and on, and their effects are as devastating to the environment as they are to our economy: chestnut blight, ash wood borer, Dutch elm disease, kudzu, Chinese privet, hemlock woody adelgid, Asian tiger mosquito, amphibian chytrid fungus, bat white-nose syndrome, rats, cats, mongoose, pigs, goats, Africanized killer bees, grass carp, Asian carp, Asian clams, giant snakeheads, swamp crayfish, lionfish, zebra mussels, lamprey, bullfrogs, nutria, fire ants, cane toads, starlings, brown trout, tilapia, Nile perch, and now Burmese pythons.

While explaining the scope of the problem, Skip evoked high-order mathematics. He asked if I'd ever heard of nondeterministic polynomial time, which he described as "problems too ridiculous to solve." The amount of time, energy, and money required to eliminate pythons in Florida is beyond imagining; it is on the same scale as controlling the Mississippi River, or channelizing, draining, and shutting off the Everglades, then changing our minds and re-creating them again. As an aside, I asked Skip about the Comprehensive Everglades Restoration Plan, also known as CERP. I honestly didn't know what was happening with the plan, and the last time I visited the Everglades it had just been initiated. He shook his head and simply said, "Next question."

CERP has barely begun to address the problems in the Everglades. Most of the attempts to reestablish flow to the surviving portions of the Everglades are a pittance. It always seems like cities, industry, and agricultural interests get first dibs. This year, the water from Lake Okeechobee is causing dramatic fish kills. And now global climate change may nullify the whole project; rising sea levels will cause the Everglades to disappear in coming decades.

Preventing establishment is the easiest and most cost-effective way to limit the effects of invasive species. Once exotics are established, a concerted effort can prevent their further spread. But once they begin expanding their range, there is little hope for eradicating them. We must remain vigilant and institute programs to prevent establishment of invasive species. We must prevent the spread of the brown treesnake to Hawaii. We must prevent the introduction of additional exotics into the United States. Unfortunately, people rarely react to a situation until it has become a crisis, and by the time a species reaches the exponential growth phase, it is usually too late to do anything about it.

The Everglades continue to die, caught between the lethal forces of rising sea levels and human water-control projects. And now a giant snake is upending the ecosystem. It's python country now.

11 · Snake Conservation

Nevertheless, again and again, in season and out of season, the question comes up, "What are rattlesnakes good for?" As if nothing that does not obviously make for the benefit of man had any right to exist; as if our ways were God's ways. Long ago an Indian to whom a French traveler put this old question replied that their tails were good for toothache, and their heads for fever. Anyhow, they are all, head and tail, good for themselves, and we need not begrudge them their share of life.

JOHN MUIR, *The Yellowstone National Park*

Wilderness Is Where You Watch Your Step

Our first campsite in Yellowstone was in a gloomy lodgepole pine forest. The backcountry ranger told us that grizzlies were in the area, and one had recently even investigated somebody's tent. He sent us there because we wanted to see grizzlies. After we made a backpacker's meal of noodles, we walked around looking for sign of them. The trunks of the nearby lodgepoles were scratched with deep grooves bleeding sap, and there were tufts of coarse reddish hair snared in the thin exfoliating bark. That night there was a gentle rain that masked the sounds outside. Still: every snap of a twig was an 800-pound grizzly bear. Every time my eyes flew open, I was certain the pitter-patter of rain on the tent fly were the footfalls of bear. Usually night sounds are a delight and are among the many reasons you go camping. That night they were a menace. In non-wilderness places, such as the woods of New York, Tennessee, Georgia, Texas, California, and eastern Kansas—anywhere without grizzlies—night sounds can be immediately dismissed as, at worst, a raccoon. In grizzly country, the sounds all seem to indicate only one thing, and a good night's sleep is a difficult proposition.

The how-to book on bears is still being written, and the rules change depending on who you talk to. I would later read that one of the worst things you can do is camp in a location where grizzlies have been marking their territory and investigating tents. Doug Peacock, a Vietnam veteran who spent decades observing wild grizzlies to recuperate from his war experiences, never camped in designated sites when bears were around. Such bears are not scared of people and far more likely to attack. Instead, he would barricade himself in thick willow jungles that bears find difficult to negotiate. Most literature on grizzlies says you shouldn't try to out climb grizzlies, but Peacock did so routinely. If the bear is just bluff-charging you, you should stand your ground and never run. If the bear is trying to defend its cubs, immediately hit the ground, assume a submissive fetal position, and protect your head and neck with your hands. But if the bear is trying to eat you, fight for your life. You practically

A prairie rattlesnake in its unmistakable defensive pose. *Photograph by Todd Pierson*

need a PhD in bear behavior to prepare for every circumstance, and I've never read anything that explained to my satisfaction how to know when a bear is trying to eat you. And it is just as well; Yellowstone should never be a sanitized, predictable place where you are completely safe. Peacock defined wilderness as "a place and only a place where one enjoys the opportunity of being attacked by a dangerous wild animal." National parks are the people's place to die like nature intended if it is the death they earned. It's a basic freedom of all Americans.

The next day, though sleep deprived, I was alive, and my mind had already scaled a new plane of alertness. We walked down trails the ranger assured us were used for travel by bears, and my brain did not slip into daydreaming. I could hear better and watched in earnest as the trail wandered through meadows of burned-out lodgepoles replaced by grasses, fireweed, and columbine. Western tanagers perched atop snags did not escape my attention. We watched a family of coyotes playing on a gravel bar along the sparkling clear brown water of a river, tumbling like Aldo Leopold's wolf family before he shot them. These coyotes would not have to fear me, but were living in fear of the wolves reintroduced to bring balance back to these mountains. The fierce green fire was burning bright again, and coyotes would need their own new sense of awareness, fear, and consciousness to remain part of this wilderness.

Our second night, we arrived in one of those rich North Woods meadows struck through with a creek, and instead of starting a fire, I chased wandering gartersnakes through the grass. They are simple, bright-eyed snakes often found in decidedly high country of the Rockies. They squirmed in bunches and painted my hands with musk when I caught a few for identification. Meanwhile my buddy strung up our food in a graceful lodgepole.

A big afternoon thunderstorm arrived from nowhere and swamped our plans for an easy dinner. Daring the lightning, we carved out a little cave under the lodgepole and dug through the wet top layer to needles and pine cones for kindling. After sculpting a meticulous cone of light materials, we had a little tongue of flame. Thick drops of rain fell. We used a poncho to guard the fire, and both of us worked with maternal diligence to blow it until the wet sticks got going. Soon it was too strong for the storm to drown, and we had the fire bellowing steam and coals to cook our dinner. We sat huddled next to our proud creation as the storm slithered off over the hills to the east. Such were the lengths to which we went to avoid going to sleep on our second night in grizzly country.

By our third day, sleep deprivation was taking a toll, and I can't say we were enjoying ourselves. My pack felt too heavy, and I thought I had torn something in my knee. It was getting hot during midday, and we were both grumpy from lack of sleep. I scanned the trail's horizon and curves for bears. A small animal scurried up a slope near the trail—brown, squat, with a tawny pair of flanks. I turned pale and a bright jolt flowered within my brain, a reaction far out of proportion to seeing the usual wildlife. It seemed about the size of a marmot, but I didn't know what to make of those flanks. There is only one mammal that fits the description, but I didn't think it occurred in Yellowstone anymore. When we got home, I discovered to my delight that it was one of the newest creatures to make its way back into Yellowstone National Park after a long hiatus: a wolverine.

Another paranoid, sleepless night. We decided to head back. We turned in our backcountry permit and got a new one, this time asking the ranger where we could camp and not have to worry about grizzlies. He sent us to a remote corner of the park rarely visited by hikers: the Black Canyon of the Yellowstone, a place of low elevations and high temperatures more similar to the Four Corners region than the mountain forests of northwestern Wyoming. We descended into the canyon and set up camp among pinyons and junipers. We never saw a bear up in the high country where we originally intended to find them. Instead, we saw nearly half a dozen cinnamon black bears amble right through our camp. We scrutinized them thoroughly, looking for the shoulder hump and dished face of grizzlies versus the "Roman nose" of the black bear. They were huge, exactly the color of a grizzly, and were bear enough for anyone.

We hiked up Black Canyon the next day, no more reassured by the ranger's insistence that this area was free of grizzlies than by our ability to identify them. The Black Canyon of the Yellowstone is a treasure rarely seen by visitors. The river that forms the Grand Canyon of the Yellowstone and Yellowstone Falls—the pounding waterfall careening down cliffs of sulfurous yellow, like a melted sunset painting—leaves the park heading northwest to the Missouri River. There it carves through blocky deposits of lava that long ago spread from our country's largest volcanic hotspot. The basalt deposits and cold lava flows are the dark, characteristic color of such geology.

We hiked through pinyon juniper woodlands a few hundred feet above the level of the river swirling blue below us. It was much hotter down in the canyon. The trail rounded a corner and then began scaling a boulder field. The boulders ranged in size from grapefruits to Volkswagens and were scattered for a few hundred feet up and 50 feet down to the water below. The river had good current but wasn't surging and sounded like a light summer rain. A few junipers grew within the scree, but otherwise there was no vegetation. The black rocks compounded the heat. But I felt good because I wasn't wearing the big heavy pack, which was back at camp. My knee was back to normal. With no grizzlies to worry about, I let my mind wander and sang one of my favorite songs under my breath. The trail wound among the boulders, so I spent about equal time negotiating the rocks and looking down the sheer face of the rocks at the river. I looked down into the swirling blue waters, then at the dark rocks, the bright sky, the dark green of the pinyons, down to the river, up to the rocks, down to the river.

The rattlesnake exploded into fury, like a spring doorstop that never settles.

I stopped in my tracks, eyes up, my hike sentenced to temporary intermission by the 2-foot prairie rattlesnake in my path. The yellow-brown rattlesnake was coiled about 5 feet ahead on the trail, taunting me with its tongue and tail. Its tongue dangled in a slow loop, then twitched three or four times rapidly before disappearing into its mouth. The snake's eyes and mine were locked. I calmed and admired it for a moment. I listened to the sustained buzzing of the rattlesnake, watching its shifting, enraged posture. It was the first time I had ever walked upon a rattlesnake without seeing it first. It warned me of the approaching danger, and for a moment it startled me out of my mind. The rattlesnake's message was clear and given without any of the ambiguity and nuance of grizzly behavior.

I've still never seen a grizzly bear and I never want to. True wilderness is their place, and I'd rather leave it to them out of respect.

Yellowstone, by anyone's definition, is a wilderness. Despite its highways and traffic jams (rangers call them "elk jams" and "bear jams"), you can hike a mile down a trail and become completely enveloped by wilderness. Edward Abbey wrote, "No doubt about it, the presence of bear, especially grizzly bear, adds a spicy titillation to a stroll in the woods." I experienced this in Yellowstone, and I've never experienced anything quite like it anywhere else in America. The canyon country, with its isolating deserts and retiring mountain lions, doesn't quite live up to the reputation. The remotest Appalachians, sanitized of large animals and hemmed on all sides by shopping malls, are not wild enough. The Okefenokee Swamp and Everglades make a strong case, but that's only if you watch too many TV shows and movies about mythical alligators. No, grizzly country is special, even if fatal grizzly attacks are even rarer than fatal snakebites.

We walked out there along the bare grassy hills and saw large animals roaming the divides on equal footing. We were just another link in the food chain, at the mercy of a larger, more accomplished predator's appetite. From a distance, we were just another animal making our way.

You also watch your step in rattlesnake country. It's not the same as grizzly country, because snakes will never go out of their way to attack you. But a trail winding around a rocky slope may bring what for many is a terrifying surprise: the enraged coils of a rattler. I don't even remember when I learned it, but I have been watching my step since I was a boy. "Watch where you put your feet." I never lose my footing on any trail, because my eyes are always scanning the trail and my brain planning my next step. Walking through southern swamps, there could be a cottonmouth anywhere. President Jimmy Carter also grew up in Georgia and knew this well, writing, "All of us learned quite early in life to be vigilant in the fields and along the streams, constantly looking before each step, so that it became a lifetime habit almost like breathing." Cottonmouths often betray their presence by flashing their namesake gums, so if you're looking out for them, you'll be fine. You tip-toe around, carefully placing each foot. Walking along a blackwater marsh in the Catskills of New York, you can wade confidently and march forth brazenly without looking. On my first trip to the tropics, it was like starting out all over again: the dreaded fer-de-lance snake was lurking, and I was out of my element. I watched my step with more scrutiny than I had in years.

I suspect that for most people, snakes evoke a fear on par with facing a wild grizzly, which, although misguided, means our country is a vast wilderness approaching Abbey and Peacock's definitions. These snakes are a pulse-quickening hazard to your strolls in the deserts, mountains, and prairies of our country, effectively expanding our American wilderness by millions of square miles. This country wouldn't be the same without them. It would be like the NFL going to flag-football rules. Like all such wilderness, these places and the snakes that make them special are under threat.

Threats to American Snakes

The number one threat to the continued existence of most American snakes is habitat loss. Habit loss is the biggest threat to all American wildlife. All the snakes that are protected by the Endangered Species Act (ESA) are on that list primarily because their habitat is fragmented and destroyed, and they are now restricted to small remnants of their formerly widespread distribution. The eastern indigo snake requires large tracts of longleaf pine sandhills with plentiful gopher tortoise burrows, along with good wetlands for foraging. These habitats were once common throughout Georgia and Florida, but they are now found only as tiny remnants scattered within a sea of agriculture and pine plantations. The copperbelly watersnake ranges across forests and marshes in the Midwest. The dark, boggy soils ringing the Great Lakes were once a vast wilderness of wet woods, swamps, and fens. Unfortunately, these soils are also terrific for growing corn, so most of the Midwest has been drained, cleared, and planted, and copperbelly watersnakes, massasaugas, and Kirtland's snakes—all rare species tied to these habitats—are now restricted to tiny, isolated pockets.

Introduced species also threaten American snakes. Perhaps the most insidious creature that doesn't belong in America is the beloved house cat. Feral cats and cats that go outdoors are rapacious killers of mice, birds, and small vertebrates like lizards and snakes. For most people, a cat bringing home a headless and broken snake is probably a cause for wonder and celebration. But cats kill an estimated 240 million native birds annually and are known to have

House cats are cute and charismatic killers of our native wildlife, including snakes. *Photograph by Zack West*

killed no fewer than 22 different kinds of native snakes. Because pet cats are well fed, they are mostly killing to indulge their instincts. They do it for fun. In Australia, feral cats are responsible for spectacular declines in their native fauna of marsupials and reptiles, and the government is currently mounting a massive cat eradication program. Australians—who love and respect their wondrous native wildlife—are all for it. In America, uninformed cat lovers and animal rights activists react in horror to such proposals, and instead advocate "spay and release" programs. The problem with spay and release is that it doesn't work, and the cats are released to kill again. I am a cat lover, and have two precious kitties of my own. One is snuggled next to me right now. I love them, which is why I keep them entirely indoors, where they are safe and have long life spans and have no opportunity to kill native wildlife. I'm an animal lover first, cat lover second.

Fortunately, other invasive species do not have a large, vocal group of champions. These include crayfish, bullfrogs, and fire ants. Large swamp crayfish and bullfrogs are now established in wetlands of the Southwest and California, where they are not native (both species are native to the eastern United States) and have been tied to declines in native species, including snakes. The crayfish is an opportunistic scavenger and forager on aquatic eggs and tadpoles, and it upends aquatic ecosystems. Bullfrogs can competitively squeeze out native amphibians and are large enough to eat most small snakes and the young of larger species. They are known predators on some of the distinctive California wetland snakes. Fire ants have spread throughout the South and are one of the most economically damaging invasive species ever established in the United States. Their venomous stings and large numbers make them effective predators on ground-nesting birds, their eggs, and small vertebrates. They cause livestock losses—fire ants have killed calves—and they have even killed people. Their arrival and spread have been linked to declines in many snakes in the South, especially the rare southern hog-nosed snake. This slow-moving snake burrows in the sand and is probably stung to death by foraging fire-ant soldiers.

Of course, some snakes are themselves invasive species, as we saw in chapter 10. Invasive species are a growing threat, becoming more and more prevalent, and Americans need to show concern and vigilance about their dangers to our native wildlife. Eradication programs for such species have so far been largely ineffective, so the focus must shift to preventing their establishment in the first place.

Emerging infectious diseases, such as HIV and Ebola, are a terrifying new class of threats to humans and wildlife. They are viruses that have recently "jumped" from other species and are now spreading in human populations. Amphibians are declining worldwide due to a fungal pathogen that hybridized within amphibian populations and spread through the international pet trade. White-nose syndrome causes collapses in bat populations and probably origi-

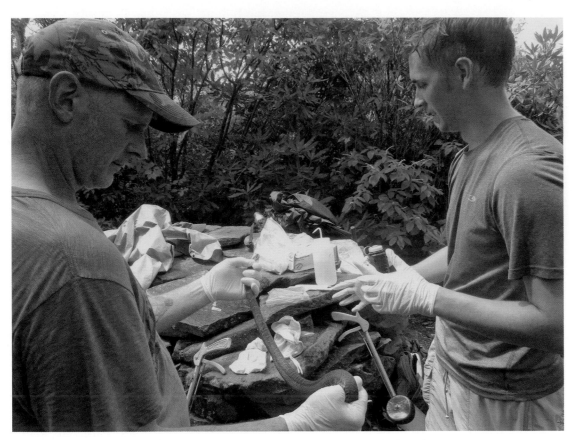

Scientists examine a snake in the Great Smoky Mountains for evidence of snake fungal disease. *Photograph by the author*

nated in Europe. All these diseases are symptomatic of our modern capacity to bring into contact species that once were isolated and confined to remote locations. Often the first sign of such diseases are mysterious declines that cannot be linked to any of the usual threats. Declines of eastern kingsnakes fit this pattern; they have simply disappeared from areas where they were once common. So far, their disappearance has not been linked to any pathogen. Alarming reports of skin lesions in snakes throughout the eastern United States have so far not been linked to any declines. The lesions are caused by yet another new fungal pathogen, and in this age of terrifying new diseases, it is cause for great concern.

Humans affect snake populations not only by destroying their habitat, but also by directly killing them, mostly with cars. Our country is now mercilessly carved up with highways of all sizes and volumes, and millions of snakes and other wildlife are killed on them every year. The roads further isolate and fragment the remaining habitat islands. Many snakes will not cross roads and avoid them, but the carnage is on a scale difficult to comprehend. Roads kill snakes with such efficiency that it is not necessary to kill snakes to collect them as museum specimens. I go out collecting snakes guilt-free; cars get dozens of them every night, and I just clean up after them.

In addition to this unintentional killing with cars (though many people do intentionally swerve to hit snakes with their cars), snakes are one of the few remaining American wildlife groups targeted for intentional, malicious killing. No other animal has experienced the level of persecution in America than snakes, and that is saying quite a lot. Snakes have long been reviled and placed in the same category as vermin that are better off dead. Few other animals receive a death sentence at the hands of an average person just for being on their property. Timber rattlesnakes were exterminated in the Appalachians and hardwood forests of the Northeast. Eastern diamondback rattlesnakes are killed on sight anywhere in the South, and are rounded up by the hundreds by professional catchers; their last stronghold will be the uninhabited barrier islands off the southern Atlantic Coast. Perhaps some of our southwestern deserts will remain a wilderness for snakes. But by stealing entire rivers, we have managed to build cities and settle millions of people even there. Millions of snake killers.

Uncountable numbers of snakes are killed by vehicular traffic each year. *Photograph by Noah Fields*

The Psychology of Fear and Hate

What is it about these predators that makes us hate them? For grizzlies, wolves, and mountain lions, it is as though these powerful hunters rival our domination over the Earth. They compete with us for game and occasionally kill our livestock. Simply put, these predators—snakes included—humble us. And if there is one thing human beings cannot abide, it is humility. That feeling you get when walking in wilderness—the anxiety that drove us from the grizzly country and back to our car—is the purest form of humility. That tip-toeing you do in rattlesnake country is an affront to your sense of power. The rattlesnake stands coiled in infuriated defiance of our manifest destiny, Pax Americana, and American exceptionalism. The Founding Fathers interpreted their stance well, using the rattlesnake as a symbol of defiance against Great Britain. But the rattlesnake doesn't choose sides in human affairs. The snake is on its own side.

Far beyond the loathing people have for the simpler annoyances caused by large predators, snakes are singled out by humans for a special kind of hatred. This is not entirely our fault. It turns out we were all born with a special capacity to fear and hate snakes. Although some cultures hold snakes in high esteem, their admiration is usually mixed with a respectful fear. Native Americans had a healthy outlook about snakes, and rattlesnakes in particular were revered and left alone by many tribes. More generally, the fear of snakes is almost entirely cross-cultural, an indication that there is a biological underpinning for it. Snakes are among the only animals that people around the world commonly dream about, and most of these dreams aren't pleasant. Even my snake dreams usually involve me being bitten by venomous snakes. Harry Greene proposed that humans have an ancestral ambivalence toward snakes; while most cultures fear them, most also recognize their value as an easy source of protein and feed on them. Taking the idea further, Greene points out that all primates—from tarsiers and lemurs to chimpanzees and people—are known to feed on and be fed upon by snakes. Others have suggested that, throughout our evolutionary history, snakes were a hazard and dreaded enemy, so naturally humans have an instinctual fear of snakes. This isn't entirely true. If ever there was an ancient enemy of primates, it was cats. Wild cats of all sorts feed on primates with great athleticism and relish, including humans. Many of our early human ancestors found their way into the fossil record after being cached by a leopard. Historically, tigers were among the few animals that added humans to the menu with regularity. Yet we keep cats in our houses, and cute cat videos dominate the Internet.

And it is not accurate to say that people have an instinctual fear of snakes. Instead, psychologists have determined that we have an innate *capacity* to fear snakes. But that fear is still learned. This can be a simple lesson taught at an early age. Even slight hesitation by an adult near a snake-like object witnessed by a child is enough to prepare the developing human brain for a lifetime of snake hating. There are few things that humans can learn so quickly at so early an age. Psychologists have identified only a few so-called nonassociative fears, including snakes and strangers. Interestingly, our primitive capacity to hate snakes is similar to our related aptitude for hating people.

If you think back far enough, you can probably remember your first encounter with a snake, when somebody either pulled you back or stepped in to kill it. I was with my brothers when we found my first snake, a handsome black ratsnake that we promptly beat to death with sticks. We carried it all the way back from the swamp and paraded it around the neighborhood, thinking the adults would crown us as heroes. We carried the snake door to door, hoping for smiles. Instead the grownups' faces twisted, and the doors quickly closed. The snake ended up in a trashcan.

Many kids have similar stories about their first encounter with snakes, and they go on to hate snakes their whole lives. But such hatred is not impossible to unlearn.

I think we have quite miraculously turned the corner in our attitudes toward snakes. It turns out that the best weapon against the fear of snakes is familiarity. People who spend time outside eventually figure out that snakes are just another kind of animal and that they aren't out to get them. People who are less familiar with snakes are the ones who develop strong phobias.

When I was a kid, some years after that first black ratsnake, my friend and I were the only ones in the whole school who liked snakes. Lately it seems that many kids are not afraid to hold a snake, and many of them actually love snakes. Because the natural fear of snakes can easily be overcome, there is reason to be optimistic about conservation in general, and although many people still hate and fear them, the number of people who respect snakes is steadily growing. There is encouraging news about snake conservation in the United States. Although the future of our snakes is still in the balance, and there is much more work to be done, I believe there is a still a place for them in our American wilderness.

Orianne

When doing snake programs for children, you never know if you're getting through to them. With a snake in one hand and the other free for gesturing, you try to be entertaining, funny, and exciting, but you still get the occasional kid who squirms away from the snake in repulsion, and they all just seem to want to tell you about how their dad killed a snake once. These demonstrations still do a lot of good, and many kids perhaps go the rest of their lives with a little less hesitation about snakes. If you do programs with snakes for kids, there is every reason to believe there might be some fairly immediate rewards for your efforts. But you will never give a snake talk that will do as much good for snake conservation as Bob Freer did in 2007.

Erika Nowak
Champion of Snakes

When she was 10 years old, Erika was exploring her family's farm in Upstate New York and discovered two gartersnakes entwined in a pasture. In the sunshine of a New York spring, gartersnakes are often out searching gardens and lawns for worms, and their bizarre mating rituals are sometimes on display. Fascinated, Erika watched them briefly before picking up a shovel and cutting them to pieces. She stared at the writhing fragments and then became placid and upset. Her father found her and saw what she'd done. What happened next was the difference between a lifetime of fear and hatred of snakes and an important career fighting to conserve them.

When she turned her research attention to rattlesnakes back in the early 1990s, there were few female herpetologists studying venomous species. She got warnings from park personnel and even other herpetologists about how aggressive rattlers were. Instead, she found that "everything I had been told was just so wrong." She feels a kinship with rattlesnakes because they are calm and gentle and yet so misunderstood. "This was the thing that just hooked me": like many of the snake biologists I spoke with, she sees snakes as underdogs. And Erika knows what it's like to be an underdog; she was born with juvenile rheumatoid arthritis, and has lived her whole life with pain and impaired mobility.

She initiated radiotelemetry studies on western diamondbacks and Arizona black rattlesnakes, following them out in the desert, up rocky bajadas, across sandy washes, through rugged mountain forests, and up

The narrow-headed gartersnake is perhaps America's most vulnerable snake. It is now found in only a few streams in New Mexico and Arizona. *Photograph by Pierson Hill*

sheer cliffs. She examined the effects of snakes humanely removed from national park campgrounds and visitor centers. Snakes moved long distances don't do well, and mortality is high. Those moved shorter distances have a better chance. Erika uses her studies to inform rangers and other personnel about snakes and their management. Nowak points out that in national parks, rangers are supposed to "start from the basis that they are a protected resource." Although her seminars are sometimes attended by good old boys who mutter under their breath about how they'll be damned if they'll save some snake, overall her programs have been wildly successful.

Nowak now studies endangered gartersnakes of the southwestern wetlands. The plight of the narrow-headed gartersnake—a bizarre stream-dwelling, fish-eating gartersnake that looks more like a watersnake—is grim. It is now restricted to perhaps three streams that appear to have healthy populations. That's three streams in the world. Streams where they were once common have been decimated by catastrophic wildfires. They are now among the rarest American snakes, and will perhaps be the first American snake to go extinct. But Erika will do everything she can to prevent that from happening on her watch.

The gartersnakes she killed when she was young still hang on Erika's conscience. She told me the story with evident remorse and embarrassment, and was relieved when I told her that I too killed a snake when I was young and ignorant. After her father found her there with her handiwork, he might have congratulated her, or admonished her for not killing the snakes without his help. Instead, he yelled, "What in the hell did you do that for?" She had no answer, and learned a lesson taught far too infrequently by parents. This, she says, "totally changed my perspective on snakes," and perhaps forged one of their greatest champions.

At that time, Bob did the snake shows at the Everglades Alligator Farm in Homestead, Florida. He had an indigo snake, which he told me was his "favorite to bring out to talk about." He told his audience the usual facts: the eastern indigo is America's largest snake and feeds on venomous snakes. The species is protected owing to habitat destruction, road mortality, illegal collection for the pet trade, and malicious killing. Bob expressed uncertainty about the effectiveness of these programs, adding, "a lot of times I don't know if I'm getting through at all," but he has also noticed a change in attitudes toward snakes for the better. He sometimes gets recognized by grownups who attended his talks when they were kids, and they thank him for inspiring them.

Back in 2007, Bob was showing off his indigo snake and there was a little girl out in front standing next to her father. She was enchanted by the snake. Indigos are among our handsomest snakes, with large, expressive eyes, a glossy-black girth, and a gentle yet regal demeanor. Despite their size and capacity to chew eastern diamondback rattlesnakes to death, indigos rarely bite people, even when first captured from the wild. The reddish barring on the face completes the package; they are spectacular, dignified creatures on par with any of our greatest American wildlife. The girl came under the snake's spell, and she watched its eyes as it swayed in Bob's hands. She looked up at her father and said, "I want you to help this snake like you do the tigers."

Her father was Thomas Kaplan, billionaire founder of Panthera, a worldwide organization dedicated to the conservation of big cats. Big cats are char-

An indigo snake: America's largest and perhaps most mystifying snake, and a federally threatened species.
Photograph by Pierson Hill

David Steen

The Next Big Thing

We bumped along in a big Chevy Silverado between endless columns of pines on a flat plain of sand. It was nine o'clock in the morning, but we were both sweating, the temperature nearing 90 degrees and the humidity always pushing saturation. Inside the cab it smelled like a roast beef sandwich. I kept asking him if he had a roast beef sandwich in there. We never found it.

We walked along lanes of metal flashing checking bucket traps for snakes. More often than not, the buckets were instead populated by huge wooly spiders, hideous cockroaches, springy daddy longlegs, and yellow scorpions. Lots and lots of scorpions. We met back up at the truck, and I asked Dave how many times he'd been stung. He thoughtfully considered the question and seemed to be counting in his head. Then, with his characteristic dry wit, he said in monotone: "Hundreds."

After driving around all day, I was out of water and tired. We pulled over and just admired the elegant bolls of the ancient longleaf pines stretching in all directions far out to the dark smudge of a developing late-afternoon thunderstorm. I told him anybody would be jealous of such a spectacular study site, and he'd always remember his time there. He said, "I'm paying my dues."

Because he publishes so many papers and he's from New York, Dave is sometimes accused of being a desk biologist. But he has certainly paid his dues in the field. And in just the past few years, he has shown rare potential as a scientist. If Henry Fitch is the grandfather of snake biology, and Harry Greene and Rick Shine are the current masters, then David Steen is the next big thing.

Photograph by the author

Dave reaches out to other researchers with his ideas, big ideas that could not be tackled by one research group alone. He combined his long-term data from across the Southeast to reveal the inner workings of snake communities. Where kingsnakes decline, copperheads increase. Where coachwhips and racers co-occur, racers are smaller and feed on smaller prey. When longleaf pine forests are healthy, snake communities respond. Now he's studying the response of the longleaf pine ecosystem to the return of the indigo snake. The project combines all of Dave's passions and strengths: he will look at big-picture ecological changes, and the results will have massive implications for conservation.

But that's Dave as a scientist. What's he really like? Among all the people I've written about, I know David the best, and I think it's important to portray scientists as real people. So, to me, this is Dave: he plays bass guitar left-handed. He gets a little crazy when he drinks too much, but there is always a great story about it the next day. He is ambitious yet modest. He doesn't suffer fools and can almost always judge a person's character after one meeting. I think that, probably because he never had brothers, he is a loyal friend and values loyalty in his friends. People don't always understand his sense of humor and don't always appreciate what he's trying to do. He claims his favorite album is Metallica's *Master of Puppets*, but I think he only says that because it sounds tough. He has the driest sense of humor but can be zany too. He once nearly got in a bar fight down in south Georgia because he took a cigarette from the mouth of some Billy and began smoking it right there in front of him. Many of Dave's friends consider him their best friend, but I'm not sure if he has a best friend. Dave is my best friend.

ismatic and beautiful, and although their giant home ranges and unavoidable conflicts with people make their conservation difficult, most people agree that wild tigers deserve a place to live. Snakes have a substantially worse image problem, so no billionaire philanthropist had ever bankrolled an organization to protect snakes. That is, until Orianne Kaplan, Thomas's daughter, met and fell in love with an indigo snake. Dedicated scientists and government agencies are now racing to save the indigo snake. Properties within the known range of the indigo are being purchased and are managed intensively for the snakes, and reintroduction efforts are now underway to reestablish the snake where they were formerly present. These programs have received boosted funding from the Orianne Society, the organization founded and bankrolled by Kaplan.

This is a remarkable conservation story wherein an animal's ability to charm a single child played a key role in its own survival. I tracked down Bob Freer, who was unaware of how much good he and his snake had done for conservation. He had never heard of Thomas Kaplan or the Orianne Society. When I talked with him, I told him about Orianne and the indigo snake conservation effort and he replied simply, "glad something good came out of it."

Two Roundups Down, One to Go

When I was growing up in Georgia, you had your choice of three rattlesnake roundups to attend each spring, where you could see hundreds of snakes brought in for slaughter. They were strung equidistantly across the southern half of the state, with little overlap for rural spectators coming from nearby towns. The Whigham Roundup served the southwestern corner of the state down near Tallahassee. The Fitzgerald Roundup was beyond I-75 halfway between Macon and Valdosta. And the largest was the Claxton Roundup, out in the barren agricultural country between Athens and Savannah. All these towns had little going for them after the interstate highway system marooned them and cash crops moved out to the Central Valley of California. Besides the rattlesnake roundup, for example, Claxton is also famous within Georgia for its fruitcakes. Fitzgerald is known for its wild chickens and is 10 miles down the road from Irwinville, where Union cavalry captured an absconding Jefferson Davis after the fall of Richmond. From what I can tell, Whigham has only the rattlesnakes and a whole lot of Confederate flags.

Enter my good buddy John. John Jensen was my first and most important mentor coming up. I met him when I was just out of high school, and we've been friends ever since. He is Georgia's state herpetologist—a single nongame wildlife biologist with the unenviable responsibility of looking after one of the largest assemblages of amphibians and reptiles of any state in the country. He was hired practically right out of college with only a zoology degree. I've always suspected they hired John because they assumed such a young, inexperienced kid would be easy to push over and tow the party line. In a state agency

A young girl develops her first impressions of snakes and wildlife at a rattlesnake roundup. *Photograph by Pierson Hill*

more interested in white-tailed deer numbers, quail populations, and rainbow trout hatcheries, a true conservation ethic can be a dangerous thing. If this was their aim, they did not get what they bargained for. John is easily the most ardent conservationist I've met, and is a more astute biologist than most PhDs I know. He has always riled things up with his position, using it as a platform to educate and take action. There is no clearer example than John's roundup takedowns.

He first approached the struggling Fitzgerald Roundup, which in the late 1990s was already in its death throes. John has a boyish face something like Paul McCartney's, and his Georgia accent is simultaneously folksy and authoritative. He is an easy guy for anyone to talk to, and he's a schmoozer—not in a slimy way, he's just a political animal and knows how to talk to people. He helped the organizers advertise, brought in funds and assistance from the Georgia Department of Natural Resources (GADNR), and helped convert the Fitzgerald Rattlesnake Roundup into the Fitzgerald Wild Chicken Festival. The town's feral jungle fowl—which have been running wild in Fitzgerald's streets, perching on its mailboxes, and crowing from its antebellum rooftops since the 1960s—are now at center stage. As of this writing, the event is still held annually, with such activities as the Michael Buras Memorial 5K, Gina's Kickin' Karaoke, and the grand finale Colony Bus Association Chicken Crowing Contest. The event's website does not even mention rattlesnakes.

Next, he approached the Evans County Wildlife Club, responsible for organizing the infamous Claxton Roundup. At first the organizers were hesitant, fearing government overreach and insidious environmentalism. Instead, they found John intelligent, easygoing, and sincerely interested in the long-term prospects of their town's traditional event. He orchestrated a radio and television advertising blitz, GADNR sponsorship, and brought in captive snakes. The roundup was converted into a wildlife festival where educational programs about snakes replaced the carnage. It's now called the Claxton Rattlesnake and Wildlife Festival. After its more ecofriendly rebranding, the event drew record crowds. The meeting's former organizer Bruce Purcell lauds Jensen's involvement and considers his role to have been pivotal. Purcell has been talking to the organizers of other roundups—including the notorious Sweetwater, Texas, roundup—about the benefits of conversion. The Claxton festival still includes hotdogs, popcorn, Brunswick stew, hot-boiled peanuts, a parade, and a beauty contest. But the rattlesnakes are there to educate, inspire, and fascinate, rather than die.

For his valiant and humble efforts to end the Georgia rattlesnake roundups, John was sued in court. At some point during his attempts to explain the harm of the events, John attracted the notice of a self-proclaimed snake venom expert who uses the roundup snakes. Somehow, this guy was interviewed on National Public Radio's *All Things Considered*, and they made him out to be some sort of hero. He is a crucial ally for the roundups that are still in business, and his claims about supplying venom for medical research—which, in the opinion of many experts, have never been satisfactorily documented—are practically the last reasonable justification for any roundup's continual existence. He therefore sued John for libel, claiming that John's objections were an attempt to destroy his livelihood. This man has a history of suing people who have unwittingly wronged him. He has shown up at scientific conferences touting his research (which none of us have ever seen published) and protesting injuries against his honor. Fortunately, the judge apparently took a dim view of his allegations and tossed out the case.

There is only one rattlesnake roundup left in Georgia, that in Whigham, and John is tirelessly encouraging the organizers to continue holding their popular event while simultaneously demanding a healthier respect for wildlife. You can contact the Whigham Community Club and the Grady County Chamber of Commerce if you'd like them to know what you think.

Can a Rattlesnake Gain Protection as an Endangered Species?

The New Mexico ridge-nosed rattlesnake was described rather recently, in 1974. After its discovery, careful analysis showed it could be differentiated from all other known ridge-nosed rattlers. As its scientific subspecies name *obscurus* implies, the variety lacks the attractive facial stripes of the other subspecies and has much more muted colors than the rest. The unique snake is found only

A New Mexico ridge-nosed rattlesnake, the only federally protected venomous snake in America. *Photograph by Marisa Ishimatsu*

in the cool forests of one small mountain range in northern Mexico and the adjacent Animas Range in extreme southwestern New Mexico. Because of its distinctiveness, Bill Degenhardt—professor of biology at New Mexico State—recalls that after the snake was described, "collectors were flocking in there." In 1974, no fewer than 15 collectors from six different states descended upon in the Animas Mountains to collect specimens. Charlie Painter, biologist with the New Mexico Department of Game and Fish, moved to protect them from overcollection. Painter urged the US Fish and Wildlife Service (FWS) to protect the snakes as endangered under the ESA. Within a year, the plan was approved, because the ranchers who owned most of property were sympathetic to the plight of the snake. The ranch and FWS hashed out an agreement to avoid damaging high-elevation areas where the snakes occurred. The company left alone the canyons and peaks and set them aside as nature reserves, and, according to Degenhardt, "they locked it up pretty tight." The property changed hands in 1982, but thankfully the new owners continued cooperating.

The story of the New Mexico ridge-nosed rattlesnake is rather remarkable for several reasons. First, this species was the first and is still the only venomous snake protected by the ESA. Dangerous animals are always the most controversial and hardest to protect; all it takes is concerned citizens with an irrational fear, and the government will often hesitate over the listing. Grizzly bears, gray wolves, and Florida panthers are among the "dangerous" animals that enjoy protected status, but opponents of these animals often use scaremongering to fight their protection. For the government to protect a venomous snake was quite unprecedented, helped greatly by the remote location where the snakes live. The small area is far from interstates, and the closest cities are the anonymous hamlets of Antelope Wells and Cloverdale, New Mexico. Human-rattlesnake conflicts involving this snake rarely occur, so it was much easier for the government to make its move. Finally, according to Degenhardt, because the snakes "were on private property and the landowners were somewhat interested in conservation," listing the rattlesnake as threatened "didn't seem to be a problem." He added with a laugh, "that helped." The ridge-nosed rattlesnake was therefore remarkably lucky. The remoteness of the habitat, willingness of landowners, and its threatened status saved the species; populations within the United States are stable and reproductive.

Ironically, perhaps the greatest threat to the species is not habitat destruction, but the continued threat of overzealous collectors. These are people who actually like snakes, who love snakes so much they want to keep a ridge-nosed rattlesnake in their house and admire its angular snout from behind the glass of an aquarium. Yet this snake is by far the least attractive of the ridge-nosed rattlers. The Arizona ridge-nose, with its reddish back and bold facial stripes, is drop-dead gorgeous, and yet the black market favors the faded and homely *obscurus*. The only reason why they want this snake is because it is rarer and harder to get. It is therefore possible that even a venomous snake can receive protection from the US government, and in this case the New Mexico ridge-nosed rattlesnake may owe its existence within the United States to this protection.

What are the prospects for other rare snakes becoming listed as endangered species? Several other snakes are also listed, but they are all nonvenomous species (the eastern indigo snake, San Francisco gartersnake, Lake Erie watersnake, copperbelly watersnake, and Atlantic salt marsh snake are examples). There are venomous species that have been petitioned for listing (the eastern diamondback rattlesnake and massasauga), and these petitions are obviously warranted. Both rattlers experienced drastic declines throughout their range, there are clear threats to their continued existence, and state laws have not proven effective at curbing the threats. The eastern diamondback is one of the few remaining examples of an American wildlife species being pushed to extinction by direct, unregulated, wanton killing. Both rattlers are still awaiting the government's decision. The eastern diamondback was petitioned for listing in 2011, but the FWS has not yet decided. The massasauga has been a candidate for listing since 1982. Both snakes live in heavily populated areas and frequently come into conflict with people. Their listing and protection will demand a level of dedication never before imagined from the FWS, the chronically underfunded and undermanned organization charged with protecting and recovering endangered species.

But if we can protect the hated eastern diamondback rattlesnake in the South, there is hope for all wilderness.

The eastern diamondback rattlesnake, the world's largest rattlesnake, is being wiped out in the South by wanton and unregulated killing. *Photograph by Daniel Wakefield*

Epilogue

RATTLESNAKE, *n.* Our prostrate brother, *Homo ventrambulans.*

AMBROSE BIERCE, *The Devil's Dictionary*

From the interdune meadows of Sapelo Island, you can hear the gentle pounding of the Atlantic Ocean just 300 yards away. The primary dunes, sparsely dressed with sea oats, are the dynamic first barricade against the waves. Here the waves are the most prominent sound, front and center of your attention; you have to talk over them. The secondary dunes are hills arrested by live oaks bearded with Spanish moss. They are more ancient and permanent. From the secondary dunes there is no sound of the ocean, and you might not even know the beach is nearby. But tucked between these are the magnificent and level interdune meadows—a carpet of thick muhly grass, light green and straw thin. The meadows are always still, a humid air hanging on them like a blanket, and in the morning the muhly grass is covered with fine dew. There is an ambient drone of high-pitched insects above the steady heartbeat of the waves. The meadows are dotted with wax myrtle and clear, shallow rainwater ponds. They are veined with a network of perfectly round tunnels made by marsh rabbits. And where there are marsh rabbits, there are diamondbacks.

To walk through these meadows, it's easiest to follow the rabbit trails. I made a permanent trail through the muhly grass by appropriating them. I made my way through the meadows to check several drift fences, walking slowly and carefully, checking not only where I put each step, but also a sizeable neighborhood around each step.

I once found a huge diamondback. Its 4-inch head was perpendicular to the trail, the first half of its body loosely coiled into an "S," the classic ambush posture of pitvipers. I don't know how long the snake was, but it is still among the largest snakes I've ever seen in the wild. Its coils were thick, and the whole snake would have fit comfortably coiled on a manhole cover. The pattern of amber-yellow diamonds and horn-brown scales virtually obliterated its outline in the grass. Its eyes were a rich hazel, the pupils narrow slits in the morning sun. The scales over the eyes gave it an angry, determined expression, but this is pure anthropomorphism; snakes cannot change facial expressions, nor can they even be said to have them. It lay there coiled, still, invisible. I stood still watching it. It watched me back, immobile, expressionless. Not even its tongue moved. I stood in the trail about 4 feet away, looking down on it. It lay in a coil, looking out. I'm sure it knew I was there. If I continued walking, my right foot would land within inches of its substantial mouth. I had steel snake tongs, which I used to tap among the grass as I walked, hoping to stir up and evict hidden snakes. I placed the tongs along my leg to block a potential strike, and walked right past the snake. I watched it as I went by, and the snake never moved. I turned around and it was still there, confident in its camouflage, patiently waiting for a marsh rabbit.

Sapelo is the third-largest barrier island along the Georgia coast, which is graced with several large islands that miraculously evaded development; three of the best, including Sapelo, are reached only by boat. Most of Sapelo Island is owned by the state and managed by the Department of Natural Resources. There is a small marine institute at the southern tip of the island owned by the University of Georgia. And there is Hog Hammock, a tiny hamlet of the Gullah people. The Gullah are direct descendants of slaves abandoned on the Sea Islands after the Civil War. Surviving there in isolation, the Gullah maintained the closest approximation to West African culture in North America.

I knew Tracy Walker, a shopkeeper who grew up in Hog Hammock, and always tried to patronize his store when I went down for research trips. Tracy left the island when he was a young man, got a college education, and returned to his home to run the store. He was a pillar of the community. I'd stop in after a sweaty day in the field, grab an ice cream, and ask Tracy how things were going. He was always so pleasant, his voice a lovely baritone, his smile big and genuine. I think we became friends right away when I recognized portraits of Booker T. Washington and W. E. B. DuBois on his shop wall. I told him I thought they were both great men and favored DuBois. He seemed interested in my research and always asked how it was going.

One day I dropped by the store, and on my way through the door I heard the unmistakable papery buzzing of a rattlesnake. I followed the sound over to the corpse of a recently decapitated diamondback, which was still writhing and rattling next to its own severed head. The shovel was leaning against Tracy's store.

Tracy came out and could see immediately that I was upset. He looked concerned and regretful.

"Now, Sean, if there was some way to let you know the snake was here, I'd have given it to you to handle it," he closed his hands and slid his right hand horizontally like a cashed check. "Flat out."

"But why'd you have to kill it?"

"There's people who come to my store. They depend on me. They're scared of snakes, and I have to protect them."

I told him I understood. The Gullah rely on Tracy for his intelligence and education. They rely on him for cokes, canned soup, boxed rice meals, and ice creams. If there's a storm coming, they know they can go to Tracy's store for batteries, bottled water, and news from the mainland. If a venomous snake was found in Hog Hammock, he was their St. George and he could slay it. Though saddened by the snake's death, I told him I didn't blame him one bit for killing it. Sapelo Islanders will not drive this magnificent snake to extinction. In fact, that beheaded snake is a counterintuitive symbol of hope. At least Tracy felt bad for killing it. But the irony of the snake and its slayer's parallel circumstances bends me backward every time I think about it.

Hundreds more snakes die annually at the hands of irrational homeowners who think they are protecting their family. They don't realize that getting close to snakes immediately puts them in more danger. Millions more snakes die on our roads every year. Their habitat shrinks ever smaller as development encroaches all natural habitats in the United States. Thousands are "rounded up" and killed en masse for the leather trade and for the morbid enjoyment of spectators. Yet the world's largest rattlesnake still survives in America. Even on the mainland, they hang on in pastures and along the edges of pine plantations and under old farm equipment. They still crawl into gopher tortoise burrows when fires burn the undergrowth. They remain in the last remnants of the once-extensive longleaf pine savannah. Their numbers remain strong, and are a common sight on the barrier islands flanking the Atlantic Coast from South Carolina to Florida, where a forgotten culture also presses on.

On Sapelo Island, they remain defiant, waiting patiently and invisible.

An eastern diamondback rattlesnake watching from its grassy kingdom. *Photograph by Daniel Wakefield*

References

Chapter 1

Abbey, E. (1982). Down the river. Dutton.

Alfaro, M. E., & Arnold, S. J. (2001). Molecular systematics and evolution of *Regina* and the Thamnophiine snakes. *Molecular Phylogenetics and Evolution* 21(3): 408–423.

Ashton, K. G., & Queiroz, A. D. (2001). Molecular systematics of the western rattlesnake, *Crotalus viridis* (Viperidae), with comments on the utility of the D-loop in phylogenetic studies of snakes. *Molecular Phylogenetics and Evolution* 21(2): 176–189.

Burbrink, F. T., & Lawson, R. (2007). How and when did Old World ratsnakes disperse into the New World? *Molecular Phylogenetics and Evolution* 43(1): 173–189.

Burbrink, F. T., Lawson, R., & Slowinski, J. B. (2000). Mitochondrial DNA phylogeography of the polytypic North American rat snake (*Elaphe obsoleta*): a critique of the subspecies concept. *Evolution* 54(6): 2107–2118.

Burke, A. C., Nelson, C. E., Morgan, B. A., & Tabin, C. (1995). Hox genes and the evolution of vertebrate axial morphology. *Development* 121(2): 333–346.

Caldwell, M. W., & Lee, M. S. (1997). A snake with legs from the marine Cretaceous of the Middle East. *Nature* 386: 705–709.

Caprette, C. L., Lee, M. S., Shine, R., Mokany, A., & Downhower, J. F. (2004). The origin of snakes (Serpentes) as seen through eye anatomy. *Biological Journal of the Linnean Society* 81(4): 469–482.

Currier, M. J. P. (1983). *Felis concolor. Mammalian Species Archive* 200: 1–7.

De Queiroz, A., Lawson, R., & Lemos-Espinal, J. A. (2002). Phylogenetic relationships of North American garter snakes (*Thamnophis*) based on four mitochondrial genes: how much DNA sequence is enough? *Molecular Phylogenetics and Evolution* 22(2): 315–329.

Dolnick, E. (2009). Down the great unknown: John Wesley Powell's 1869 journey of discovery and tragedy through the Grand Canyon. HarperCollins.

Douglas, M. E., Douglas, M. R., Schuett, G. W., & Porras, L. W. (2006). Evolution of rattlesnakes (Viperidae; *Crotalus*) in the warm deserts of western North America shaped by Neogene vicariance and Quaternary climate change. *Molecular Ecology* 15(11): 3353–3374.

Douglas, M. E., Douglas, M. R., Schuett, G. W., Porras, L. W., & Holycross, A. T. (2002). Phylogeography of the western rattlesnake (*Crotalus viridis*) complex, with emphasis on the Colorado Plateau. Pp. 11–50. In: Biology of the vipers. M. E. Douglas and G. W. Shuett (eds). Eagle Mountain.

Goris, R., Nakano, M., Atobe, Y., Kadota, T., Funakoshi, K., Hisajima, T., & Kishida, R. (2000). Nervous control of blood flow microkinetics in the infrared organs of pit vipers. *Autonomic Neuroscience* 84(1): 98–106.

Greene, H. W. (1983). Dietary correlates of the origin and radiation of snakes. *American Zoologist* 23(2): 431–441.

Greene, H. W. (1992). The ecological and behavioral context for pitviper evolution. Pp. 107–117. In: Biology of the pitvipers. J. Campbell and E. Brodie Jr. (eds.). Selva.

Greene, H. W. (1997). Snakes: the evolution of mystery in nature. University of California Press.

Greer, A. E. (1991). Limb reduction in squamates: identification of the lineages and discussion of the trends. *Journal of Herpetology* 25: 166–173.

Hay, O. P. 1887. The massasauga and its habits. *American Naturalist* 21: 211–218.

Lee, M. S. (2005). Molecular evidence and marine snake origins. *Biology Letters* 1(2): 227–230.

Longrich, N. R., Bhullar, B. A. S., & Gauthier, J. A. (2012). A transitional snake from the Late Cretaceous period of North America. *Nature* 488(7410): 205–208.

McBride, R., & McBride, C. (2010). Predation of a large alligator by a Florida panther. *Southeastern Naturalist* 9(4): 854–856.

Molenaar, G. J. (1992). Anatomy and physiology of infrared sensitivity in snakes. Pp. 367–454. In: Biology of the reptilia, vol. 17. C. Gans and P. S. Ulinski (eds.). University of Chicago Press.

Powell, J. W. (2012). The exploration of the Colorado River and its canyons. Courier Dover.

Pyron, R. A., & Burbrink, F. T. (2009). Systematics of the common kingsnake (*Lampropeltis getula*; Serpentes: Colubridae) and the burden of heritage in taxonomy. *Zootaxa* 2241: 22–32.

Pyron, R. A., Burbrink, F. T., Colli, G. R., De Oca, A. N. M., Vitt, L. J., Kuczynski, C. A., & Wiens, J. J. (2011). The phylogeny of advanced snakes (Colubroidea), with discovery of a new subfamily and comparison of support methods for likelihood trees. *Molecular Phylogenetics and Evolution* 58(2): 329–342.

Pyron, R. A., Burbrink, F. T., & Wiens, J. J. (2013). A phylogeny and revised classification of Squamata, including 4161 species of lizards and snakes. *BMC Evolutionary Biology* 13(1): 1.

Rodríguez-Robles, J. A., & De Jesús-Escobar, J. M. (2000). Molecular systematics of new world gopher, bull, and pinesnakes (*Pituophis*: Colubridae), a transcontinental species complex. *Molecular Phylogenetics and Evolution* 14(1): 35–50.

Schuett, G. W., D. L. Clark, & Kraus, F. 1984. Feeding mimicry in the rattlesnake *Sistrurus catenatus* with comments on the evolution of the rattle. *Animal Behaviour* 32: 625–626.

Sisk, N. R., & Jackson, J. F. (1997). Tests of two hypotheses for the origin of the crotaline rattle. *Copeia* 1997: 485–495.

Vidal, N., & Hedges, S. B. (2004). Molecular evidence for a terrestrial origin of snakes. *Proceedings of the Royal Society of London B: Biological Sciences* 271: 226–229.

Wallace, D. R. (2000). The bonehunters' revenge: dinosaurs, greed, and the greatest scientific feud of the Gilded Age. Houghton Mifflin Harcourt.

Williams, G. C. (1966). Adaptation and natural selection. Princeton University Press.

Chapter 2

Amemiya, F., Nakano, M., Goris, R. C., Kadota, T., Atobe, Y., Funakoshi, K., & Kishida, R. (1999). Microvasculature of crotaline snake pit organs: possible function as a heat exchange mechanism. *Anatomical Record* 254(1): 107–115.

Bennett, A. F. (1994). Exercise performance of reptiles. *Advances in Veterinary Science and Comparative Medicine* 38: 113–113.

Cundall, D. 1987. Functional morphology. Pp. 106–140. In: Snakes: ecology and evolutionary biology. R. A. Seigel, J. T. Collins, and S. S. Novak (eds.). Macmillan.

Filoramo, N. I., & Schwenk, K. (2009). The mechanism of chemical delivery to the vomeronasalorgans in squamate reptiles: a comparative morphological approach. *Journal of Experimental Zoology Part A: Ecological Genetics and Physiology* 311(1): 20–34.

Gans, C. (1986). Locomotion of limbless vertebrates: pattern and evolution. *Herpetologica* 42(1): 33–46.

Gove, D., & Burghardt, G. M. (1983). Context-correlated parameters of snake and lizard tongue-flicking. *Animal Behaviour* 31(3): 718–723.

Greene, H. W., Fogden, M., & Fogden, P. (2000). Snakes: the evolution of mystery in nature. University of California Press.

Greenwald, O. E. (1971). The effect of body temperature on oxygen consumption and heart rate in the Sonora gopher snake, *Pituophis catenifer affinis* Hallowell. *Copeia* 1971: 98–106.

Halpern, M. (1987). The organization and function of the vomeronasal system. *Annual Review of Neuroscience* 10(1): 325–362.

Halpern, M. (1992). Nasal chemical senses in reptiles: structure and function. *Biology of the Reptilia* 18: 423–523.

Hill, R. W., Wyse, G. A., & Anderson, M. 2012. Animal physiology, 3rd ed. Sinauer Associates.

Jayne, B. C. (1988). Muscular mechanisms of snake locomotion: an electromyographic study of the sidewinding and concertina modes of *Crotalus cerastes*, *Nerodia fasciata* and *Elaphe obsoleta*. *Journal of Experimental Biology* 140(1): 1–33.

Jonas, J. B., Schneider, U., & Naumann, G. O. (1992). Count and density of human retinal photoreceptors. *Graefe's Archive for Clinical and Experimental Ophthalmology* 230(6): 505–510.

Kubie, J. L., Vagvolgyi, A., & Halpern, M. (1978). Roles of the vomeronasal and olfactory systems in courtship behavior of male garter snakes. *Journal of Comparative and Physiological Psychology* 92(4): 627.

Lillywhite, H. B. (2014). How snakes work: structure, function and behavior of the world's snakes. Oxford University Press.

Lillywhite, H. B., de Delva, P., & Noonan, B. P. (2002). Patterns of gut passage time and the chronic retention of fecal mass in viperid snakes. Pp. 497–506. In Biology of the vipers. M. E. Douglas and G. W. Shuett (eds). Eagle Mountain.

Marvi, H., & Hu, D. L. (2012). Friction enhancement in concertina locomotion of snakes. *Journal of the Royal Society Interface* 9(76): 3067–3080.

Mason, R. T. (1992). Reptilian pheromones. *Biology of the Reptilia* 18: 114–228.

Mason, R. T., & Parker, M. R. (2010). Social behavior and pheromonal communication in reptiles. *Journal of Comparative Physiology A* 196(10): 729–749.

Meredith, M. (2001). Human vomeronasal organ function: a critical review of best and worst cases. *Chemical Senses* 26(4): 433–445.

Molenaar, G. J. (1992). Anatomy and physiology of infrared sensitivity of snakes. *Biology of the Reptilia* 17: 367–453.

Nelson, R. J. (2005). An introduction to behavioral endocrinology. Sinauer Associates.

Pough, F. H. (1980). The advantages of ectothermy for tetrapods. *American Naturalist* 115(1): 92–112.

Pough, F. H. (1983). Amphibians and reptiles as low-energy systems. Pp. 141–188. In Behavioral energetics: the cost of survival in vertebrates. W. P. Aspey and S. K. Lustick (eds.). Ohio State University Press.

Pough, F. H., Janis, C. M., & Heiser, J. B. (1999). Vertebrate life, vol. 733. Prentice Hall.

Secor, S. (1992). A preliminary analysis of the movement and home range size of the sidewinder, *Crotalus cerastes*. Pp. 389–393. In Biology of the pitvipers. J. A. Campbell and E. D. Brodie Jr. (eds.). Selva.

Secor, S. M. (1994). Ecological significance of movements and activity range for the sidewinder, *Crotalus cerastes*. *Copeia* 1994: 631–645.

Secor, S. M., & Diamond, J. (1998). A vertebrate model of extreme physiological regulation. *Nature* 395(6703): 659–662.

Secor, S. M., Fehsenfeld, D., Diamond, J., & Adrian, T. E. (2001). Responses of python gastrointestinal regulatory peptides to feeding. *Proceedings of the National Academy of Sciences* 98(24): 13637–13642.

Secor, S. M., Jayne, B. C., & Bennett, A. F. (1992). Locomotor performance and energetic cost of sidewinding by the snake *Crotalus cerastes*. *Journal of Experimental Biology* 163(1): 1–14.

Secor, S. M., & Nagy, K. A. (1994). Bioenergetic correlates of foraging mode for the snakes *Crotalus cerastes* and *Masticophis flagellum*. *Ecology* 75(6): 1600–1614.

Shmida, A. (1985). Biogeography of the desert floras of the world. *Ecosystems of the World* 12: 23–77.

Sillman, A. J., Govardovskii, V. I., Röhlich, P., Southard, J. A., & Loew, E. R. (1997). The photoreceptors and visual pigments of the garter snake (*Thamnophis sirtalis*): a microspectrophotometric, scanning electron microscopic and immunocytochemical study. *Journal of Comparative Physiology A* 181(2): 89–101.

Sillman, A. J., Carver, J. K., & Loew, E. R. (1999). The photoreceptors and visual pigments in the retina of a boid snake, the ball python (*Python regius*). *Journal of Experimental Biology* 202(14): 1931–1938.

Smith, C. F., Schwenk, K., Earley, R. L., & Schuett, G. W. (2008). Sexual size dimorphism of the tongue in a North American pitviper. *Journal of Zoology* 274(4): 367–374.

Wallach, V. (1998). The lungs of snakes. *Biology of the Reptilia* 19: 93–295.

Young, B. A. (1997). On the absence of taste buds in monitor lizards (*Varanus*) and snakes. *Journal of Herpetology* 31(1): 130–137.

Young, B. A. (2003). Snake bioacoustics: toward a richer understanding of the behavioral ecology of snakes. *Quarterly Review of Biology* 78(3): 303–325.

Chapter 3

Beck, D. D. (1995). Ecology and energetics of three sympatric rattlesnake species in the Sonoran Desert. *Journal of Herpetology* 29: 211–223.

Clarke, J. A., Chopko, J. T., & Mackessy, S. P. (1996). The effect of moonlight on activity patterns of adult and juvenile prairie rattlesnakes (*Crotalus viridis viridis*). *Journal of Herpetology* 30: 192–197.

Durner, G. M., & Gates, J. E. (1993). Spatial ecology of black rat snakes on Remington Farms, Maryland. *Journal of Wildlife Management* 57: 812–826.

Duvall, D., King, M. B., & Gutzwiller, K. J. (1985). Behavioral ecology and ethology of the prairie rattlesnake. *National Geographic Research* 1(1): 80–111.

Elbroch, M., & Rinehart, K. (2011). Peterson reference guide to the behavior of North American mammals. Houghton Mifflin Harcourt.

Ernst, C. H., & Ernst, E. M. (2003). Snakes of the United States and Canada. Smithsonian Books.

Gerald, G. W., Bailey, M. A., & Holmes, J. N. (2006). Movements and activity range sizes of northern pinesnakes (*Pituophis melanoleucus melanoleucus*) in middle Tennessee. *Journal of Herpetology* 40(4): 503–510.

Hudson, C. (1998). Knights of Spain, warriors of the sun: Hernando de Soto and the South's ancient chiefdoms. University of Georgia Press.

Huey, R. B., Peterson, C. R., Arnold, S. J., & Porter, W. P. (1989). Hot rocks and not-so-hot rocks: retreat-site selection by garter snakes and its thermal consequences. *Ecology* 70(4): 931–944.

Lillywhite, H. B., & Brischoux, F. (2012). Is it better in the moonlight? Nocturnal activity of insular cottonmouth snakes increases with lunar light levels. *Journal of Zoology* 286(3): 194–199.

Macartney, J. M., Gregory, P. T., & Larsen, K. W. (1988). A tabular survey of data on movements and home ranges of snakes. *Journal of Herpetology* 22(1): 61–73.

May, P. G., Farrell, T. M., Heulett, S. T., Pilgrim, M. A., Bishop, L. A., Spence, D. J., & Richardson, W. E. (1996). Seasonal abundance and activity of a rattlesnake (*Sistrurus miliarius barbouri*) in central Florida. *Copeia* 1996: 389–401.

Nelson, K. J., & Gregory, P. T. (2000). Activity patterns of garter snakes, Thamnophis sirtalis, in relation to weather conditions at a fish hatchery on Vancouver Island, British Columbia. *Journal of Herpetology* 34(1): 32–40.

Plummer, M. V., & Congdon, J. D. (1994). Radiotelemetric study of activity and movements of racers (*Coluber constrictor*) associated with a Carolina bay in South Carolina. *Copeia* 1994: 20–26.

Pough, F. H. (1980). The advantages of ectothermy for tetrapods. *American Naturalist* 115(1): 92–112.

Pough, F. H. (1983). Amphibians and reptiles as low-energy systems. Pp. 141–188. In Behavioral energetics: the cost of survival in vertebrates. W. P. Aspey and S. K. Lustick (eds.). Ohio State University Press.

Pough, F. H., Janis, C. M., & Heiser, J. B. (1999). Vertebrate life. Prentice Hall.

Reinert, H. K., & Cundall, D. (1982). An improved surgical implantation method for radio-tracking snakes. *Copeia* 1982: 702–705.

Richardson, M. L., Weatherhead, P. J., & Brawn, J. D. (2006). Habitat use and activity of prairie kingsnakes (*Lampropeltis calligaster calligaster*) in Illinois. *Journal of Herpetology* 40(4): 423–428.

Rosen, P. C. (1991). Comparative field study of thermal preferenda in garter snakes (*Thamnophis*). *Journal of Herpetology* 25(3): 301–312.

Secor, S. M. (1995). Ecological aspects of foraging mode for the snakes *Crotalus cerastes* and *Masticophis flagellum*. *Herpetological Monographs* 9: 169–186.

Sperry, J. H., Ward, M. P., & Weatherhead, P. J. (2013). Effects of temperature, moon phase, and prey on nocturnal activity in ratsnakes: an automated telemetry study. *Journal of Herpetology* 47(1): 105–111.

Todd, B. D., Willson, J. D., Winne, C. T., & Gibbons, J. W. (2008). Aspects of the ecology of the earth snakes (*Virginia valeriae* and *V. striatula*) in the upper coastal plain. *Southeastern Naturalist* 7(2): 349–358.

Williams, D. (1994). The Georgia gold rush: Twenty-Niners, Cherokees, and gold fever. William S. Hein.

Willson, J. D., & Dorcas, M. E. (2004). Aspects of the ecology of small fossorial snakes in the western Piedmont of North Carolina. *Southeastern Naturalist* 3(1): 1–12.

Whiteman, H. H., Mills, T. M., Scott, D. E., & Gibbons, J. W. (1995). Confirmation of a range extension for the pine woods snake (*Rhadinaea flavilata*). *Herpetological Review* 26(3): 158.

Wund, M. A., Torocco, M. E., Zappalorti, R. T., & Reinert, H. K. (2007). Activity ranges and habitat use of *Lampropeltis getula getula* (Eastern Kingsnakes). *Northeastern Naturalist* 14(3): 343–360.

Chapter 4

Aldridge, R. D. (1975). Environmental control of spermatogenesis in the rattlesnake *Crotalus viridis*. *Copeia* 1975: 493–496.

Aldridge, R. D. (1993). Male reproductive anatomy and seasonal occurrence of mating and combat behavior of the rattlesnake *Crotalus v. viridis*. *Journal of Herpetology* 27(4): 481–484.

Aleksiuk, M., & Gregory, P. T. (1974). Regulation of seasonal mating behavior in *Thamnophis sirtalis parietalis*. *Copeia* 1974: 681–689.

Arndt, R. G. (1980). A hibernating eastern hognose snake (*Heterodon platyrhinos*). *Herpetological Review* 11: 30–32.

Barbour, R. W., Harvey, M. J., & Hardin, J. W. (1969). Home range, movements, and activity of the eastern worm snake, *Carphophis amoenus amoenus*. *Ecology* 50(3): 470–476.

Baxley, D., Lipps Jr., G. J., & Qualls, C. P. (2011). Multiscale habitat selection by black pine snakes (*Pituophis melanoleucus lodingi*) in southern Mississippi. *Herpetologica* 67(2): 154–166.

Beane, J. C. (2007). *Heterodon simus* (southern hognose snake). Hibernacula. *Herpetological Review* 38: 467.

Bernardino, F. S., & Dalrymple, G. H. (1992). Seasonal activity and road mortality of the snakes of the Pa-hay-okee wetlands of Everglades National Park, USA. *Biological Conservation* 62(2): 71–75.

Blem, C. R., & Blem, L. B. (1995). The eastern cottonmouth (*Agkistrodon piscivorus*) at the northern edge of its range. *Journal of Herpetology* 29(3): 391–398.

Breininger, D. R., Bolt, M. R., Legare, M. L., Drese, J. H., & Stolen, E. D. (2011). Factors influencing home-range sizes of eastern indigo snakes in central Florida. *Journal of Herpetology* 45(4): 484–490.

Brown, W. S., & MacLean, F. M. (1983). Conspecific scent-trailing by newborn timber rattlesnakes, *Crotalus horridus*. *Herpetologica* 39(4): 430–436.

Brown, W. S., & Parker, W. S. (1976). Movement ecology of *Coluber constrictor* near communal hibernacula. *Copeia* 1976: 225–242.

Burger, J., & Zappalorti, R. T. (1986). Nest site selection by pine snakes, *Pituophis melanoleucus*, in the New Jersey pine barrens. *Copeia* 1986: 116–121.

Burger, J., Zappalorti, R. T., Dowdell, J., Georgiadis, T., Hill, J., & Gochfeld, M. (1992). Subterranean predation on pine snakes (*Pituophis melanoleucus*). *Journal of Herpetology* 26(3): 259–263.

Burger, J., Zappalorti, R. T., Gochfeld, M., Boarman, W. I., Caffrey, M., Doig, V., & Saliva, J. (1988). Hibernacula and summer den sites of pine snakes (*Pituophis melanoleucus*) in the New Jersey pine barrens. *Journal of Herpetology* 22(4): 425–433.

Burger, J., Zappalorti, R. T., Gochfeld, M., DeVito, E., Schneider, D., McCort, M., & Jeitner, C. (2012). Long-term use of hibernacula by northern pinesnakes (*Pituophis melanoleucus*). *Journal of Herpetology* 46(4): 596–601.

Cardwell, M. D. (2008). The reproductive ecology of Mohave rattlesnakes. *Journal of Zoology* 274(1): 65–76.

Crews, D., Camazine, B., Diamond, M., Mason, R., Tokarz, R. R., & Garstka, W. R. (1984). Hormonal independence of courtship behavior in the male garter snake. *Hormones and Behavior* 18(1): 29–41.

Dalrymple, G. H., & Reichenbach, N. G. (1981). Interactions between the prairie garter snake (*Thamnophis radix*) and the common garter snake (*T. sirtalis*) in Killdeer Plains, Wyandot County, Ohio. *Ohio Biological Survey Biology Notes* 15: 244–250.

Dalrymple, G. H., Steiner, T. M., Nodell, R. J., & Bernardino Jr., F. S. (1991). Seasonal activity of the snakes of Long Pine Key, Everglades National Park. *Copeia* 1991: 294–302.

Duvall, D., King, M. B., & Gutzwiller, K. J. (1985). Behavioral ecology and ethology of the prairie rattlesnake. *National Geographic Research* 1(1): 80–111.

Gienger, C. M., & Beck, D. D. (2011). Northern Pacific rattlesnakes (*Crotalus oreganus*) use thermal and structural cues to choose overwintering hibernacula. *Canadian Journal of Zoology* 89(11): 1084–1090.

Gillingham, J. C., & Carpenter, C. C. (1978). Snake hibernation: construction of and observations on a man-made hibernaculum (Reptilia, Serpentes). *Journal of Herpetology* 12(4): 495–498.

Gregory, P. T. (1974). Patterns of spring emergence of the red-sided garter snake (*Thamnophis sirtalis parietalis*) in the Interlake region of Manitoba. *Canadian Journal of Zoology* 52(8): 1063–1069.

Gregory, P. T. (1982). Reptilian hibernation. *Biology of the Reptilia* 13: 53–154.

Gregory, P. T., & Stewart, K. W. (1975). Long-distance dispersal and feeding strategy of the red-sided garter snake (*Thamnophis sirtalis parietalis*) in the Interlake of Manitoba. *Canadian Journal of Zoology* 53(3): 238–245.

Harvey, D. S., & Weatherhead, P. J. (2006). Hibernation site selection by eastern massasauga rattlesnakes (*Sistrurus catenatus catenatus*) near their northern range limit. *Journal of Herpetology* 40(1): 66–73.

Huang, W. S., Greene, H. W., Chang, T. J., & Shine, R. (2011). Territorial behavior in

Taiwanese kukrisnakes (*Oligodon formosanus*). *Proceedings of the National Academy of Sciences* 108(18): 7455–7459.

Hyslop, N. L., Meyers, J. M., Cooper, R. J., & Stevenson, D. J. (2014). Effects of body size and sex of *Drymarchon couperi* (eastern indigo snake) on habitat use, movements, and home range size in Georgia. *Journal of Wildlife Management* 78(1): 101–111.

Jellen, B. C., & Aldridge, R. D. (2014). It takes two to tango: female movement facilitates male mate location in wild northern watersnakes (*Nerodia sipedon*). *Behaviour* 151(4): 421–434.

Jellen, B. C., & Kowalski, M. J. (2007). Movement and growth of neonate eastern massasaugas (*Sistrurus catenatus*). *Copeia* 2007: 994–1000.

Kapfer, J. M., Coggins, J. R., & Hay, R. (2008). Spatial ecology and habitat selection of bullsnakes (*Pituophis catenifer sayi*) at the northern periphery of their geographic range. *Copeia* 2008: 815–826.

Keller, W. L., & Heske, E. J. (2000). Habitat use by three species of snakes at the Middle Fork Fish and Wildlife Area, Illinois. *Journal of Herpetology* 34(4): 558–564.

King, M. B., & Duvall, D. (1990). Prairie rattlesnake seasonal migrations: episodes of movement, vernal foraging and sex differences. *Animal Behaviour* 39(5): 924–935.

Kingsbury, B. A., & Coppola, C. J. (2000). Hibernacula of the copperbelly water snake (*Nerodia erythrogaster neglecta*) in southern Indiana and Kentucky. *Journal of Herpetology* 34(2): 294–298.

Linehan, J. M., Smith, L. L., & Steen, D. A. (2010). Ecology of the eastern kingsnake (*Lampropeltis getula getula*) in a longleaf pine (*Pinus palustris*) forest in southwestern Georgia. *Herpetological Conservation and Biology* 5(1): 94–101.

Macartney, J. M., Gregory, P. T., & Larsen, K. W. (1988). A tabular survey of data on movements and home ranges of snakes. *Journal of Herpetology* 22(1): 61–73.

Miller, G. J., Smith, L. L., Johnson, S. A., & Franz, R. (2012). Home range size and habitat selection in the Florida pine snake (*Pituophis melanoleucus mugitus*). *Copeia* 2012: 706–713.

Mills, M. S., Hudson, C. J., & Berna, H. J. (1995). Spatial ecology and movements of the brown water snake (*Nerodia taxispilota*). *Herpetologica* 51(4): 412–423.

Mitrovich, M. J., Diffendorfer, J. E., & Fisher, R. N. (2009). Behavioral response of the coachwhip (*Masticophis flagellum*) to habitat fragment size and isolation in an urban landscape. *Journal of Herpetology* 43(4): 646–656.

Moore, J. A., & Gillingham, J. C. (2006). Spatial ecology and multi-scale habitat selection by a threatened rattlesnake: the eastern massasauga (*Sistrurus catenatus catenatus*). *Copeia* 2006: 742–751.

Parker, W. S., & Brown, W. S. (1973). Species composition and population changes in two complexes of snake hibernacula in northern Utah. *Herpetologica* 29(4): 319–326.

Pisani, G. R. (2009). Use of an active ant nest as a hibernaculum by small snake species. *Transactions of the Kansas Academy of Science* 112(1/2): 113–118.

Reinert, H. K. (1984). Habitat separation between sympatric snake populations. *Ecology* 65(2): 478–486.

Reinert, H. K., & Zappalorti, R. T. (1988a). Field observation of the association of adult and neonatal timber rattlesnakes, *Crotalus horridus*, with possible evidence for conspecific trailing. *Copeia* 1988: 1057–1059.

Reinert, H. K., & Zappalorti, R. T. (1988b). Timber rattlesnakes (*Crotalus horridus*) of the pine barrens: their movement patterns and habitat preference. *Copeia* 1988: 964–978.

Robson, L. E. (2011). The spatial ecology of eastern hognose snakes (*Heterodon platirhinos*): habitat selection, home range size, and the effect of roads on movement patterns (PhD dissertation, University of Ottawa).

Roe, J. H., Kingsbury, B. A., & Herbert, N. R. (2004). Comparative water snake ecology: conservation of mobile animals that use temporally dynamic resources. *Biological Conservation* 118(1): 79–89.

Roth, E. D., May, P. G., & Farrell, T. M. (1999). Pigmy rattlesnakes use frog-derived chemical cues to select foraging sites. *Copeia* 1999: 772–774.

Row, J. R., Blouin-Demers, G., & Lougheed, S. C. (2012). Movements and habitat use of eastern foxsnakes (*Pantherophis gloydi*) in two areas varying in size and fragmentation. *Journal of Herpetology* 46(1): 94–99.

Schuett, G. W., Repp, R. A., Taylor, E. N., DeNardo, D. F., Earley, R. L., Van Kirk, E. A., & Murdoch, W. J. (2006). Winter profile of plasma sex steroid levels in free-living male western diamond-backed rattlesnakes, *Crotalus atrox* (Serpentes: Viperidae). *General and Comparative Endocrinology* 149(1): 72–80.

Shew, J. J., Greene, B. D., & Durbian, F. E. (2012). Spatial ecology and habitat use of the western foxsnake (*Pantherophis vulpinus*) on Squaw Creek National Wildlife Refuge (Missouri). *Journal of Herpetology* 46(4): 539–548.

Sperry, J. H., Blouin-Demers, G., Carfagno, G. L., & Weatherhead, P. J. (2010). Latitudinal variation in seasonal activity and mortality in ratsnakes (*Elaphe obsoleta*). *Ecology* 91(6): 1860–1866.

Sperry, J. H., Peak, R. G., Cimprich, D. A., & Weatherhead, P. J. (2008). Snake activity affects seasonal variation in nest predation risk for birds. *Journal of Avian Biology* 39(4): 379–383.

Steen, D. A., Steen, A. D., Pokswinski, S., Graham, S. P., & Smith, L. L. (2010). Snakes using stumpholes and windfall tree-associated subterranean structures in longleaf pine forests. *Reptiles and Amphibians* 17: 49–51.

Steen, D. A., Linehan, J. M., & Smith, L. L. (2010). Multiscale habitat selection and refuge use of common kingsnakes, *Lampropeltis getula*, in southwestern Georgia. *Copeia* 2010: 227–231.

Tiebout III, H. M., & Cary, J. R. (1987). Dynamic spatial ecology of the water snake, *Nerodia sipedon*. *Copeia* 1987: 1–18.

Waldron, J. L., Lanham, J. D., & Bennett, S. H. (2006). Using behaviorally-based seasons to investigate canebrake rattlesnake (*Crotalus horridus*) movement patterns and habitat selection. *Herpetologica* 62(4): 389–398.

Wastell, A. R., & Mackessy, S. P. (2011). Spatial ecology and factors influencing movement patterns of Desert Massasauga rattlesnakes (*Sistrurus catenatus edwardsii*) in southeastern Colorado. *Copeia* 2011: 29–37.

Weatherhead, P. J., & Hoysak, D. J. (1989). Spatial and activity patterns of black rat snakes (*Elaphe obsoleta*) from radiotelemetry and recapture data. *Canadian Journal of Zoology* 67(2): 463–468.

Weatherhead, P. J., & Prior, K. A. (1992). Preliminary observations of habitat use and movements of the eastern massasauga rattlesnake (*Sistrurus c. catenatus*). *Journal of Herpetology* 26(4): 447–452.

Whittier, J. M., Mason, R. T., Crews, D., & Licht, P. (1987). Role of light and temperature in the regulation of reproduction in the red-sided garter snake, *Thamnophis sirtalis parietalis. Canadian Journal of Zoology* 65(8): 2090–2096.

Wund, M. A., Torocco, M. E., Zappalorti, R. T., & Reinert, H. K. (2007). Activity ranges and habitat use of *Lampropeltis getula getula* (Eastern Kingsnakes). *Northeastern Naturalist* 14(3): 343–360.

Chapter 5

Alcock, J. (1989). Animal behavior. Sunderland: Sinauer Associates.

Aldridge, R. D. (1982). The ovarian cycle of the watersnake *Nerodia sipedon*, and effects of hypophysectomy and gonadotropin administration. *Herpetologica* 38(1): 71–79.

Aldridge, R. D., & Bufalino, A. P. (2003). Reproductive female common watersnakes (*Nerodia sipedon sipedon*) are not anorexic in the wild. *Journal of Herpetology* 37(2): 416–419.

Aldridge, R. D., & Duvall, D. (2002). Evolution of the mating season in the pitvipers of North America. *Herpetological Monographs* 16(1): 1–25.

Barry, F. E., Weatherhead, P. J., & Philipp, D. P. (1992). Multiple paternity in a wild population of northern water snakes, *Nerodia sipedon. Behavioral Ecology and Sociobiology* 30(3–4): 193–199.

Bernhardt, P. C., Dabbs Jr., J. M., Fielden, J. A., & Lutter, C. D. (1998). Testosterone changes during vicarious experiences of winning and losing among fans at sporting events. *Physiology and Behavior* 65(1): 59–62.

Birchard, G. F., Black, C. P., Schuett, G. W., & Black, V. (1984). Foetal-maternal blood respiratory properties of an ovoviviparous snake the cottonmouth, *Agkistrodon piscivorus. Journal of Experimental Biology* 108(1): 247–255.

Birkhead, T. R., & Møller, A. P. (1993). Sexual selection and the temporal separation of reproductive events: sperm storage data from reptiles, birds and mammals. *Biological Journal of the Linnean Society* 50(4): 295–311.

Blackburn, D. G., Stewart, J. R., Baxter, D. C., & Hoffman, L. H. (2002). Placentation in garter snakes: scanning EM of the placental membranes of *Thamnophis ordinoides* and *T. sirtalis. Journal of Morphology* 252(3): 263–275.

Blouin-Demers, G., Weatherhead, P. J., & Row, J. R. (2004). Phenotypic consequences of nest-site selection in black rat snakes (*Elaphe obsoleta*). *Canadian Journal of Zoology* 82(3): 449–456.

Brown, W. S. (1991). Female reproductive ecology in a northern population of the timber rattlesnake, *Crotalus horridus. Herpetologica* 47(1): 101–115.

Carpenter, C. C., & Ferguson, G. W. (1977). Variation and evolution of stereotyped behavior in reptiles. *Biology of the Reptilia* 7: 335–554.

Clesson, D., Bautista, A., Baleckaitis, D. D., & Krohmer, R. W. (2002). Reproductive biology of male eastern garter snakes (*Thamnophis sirtalis sirtalis*) from a denning population in central Wisconsin. *American Midland Naturalist* 147(2): 376–386.

Cunnington, G. M., & Cebek, J. E. (2005). Mating and nesting behavior of the eastern hognose snake (*Heterodon platirhinos*) in the northern portion of its range. *American Midland Naturalist* 154(2): 474–478.

Duvall, D., & Schuett, G. W. (1997). Straight-line movement and competitive mate searching in prairie rattlesnakes, *Crotalus viridis viridis*. *Animal Behaviour* 54(2): 329–334.

Ernst, C. H., & Ernst, E. M. (2003). Snakes of the United States and Canada. Smithsonian Books.

Fitzgerald, L. A., & Painter, C. W. (2000). Rattlesnake commercialization: long-term trends, issues, and implications for conservation. *Wildlife Society Bulletin* 28(1): 235–253.

Ford, N. B., & Seigel, R. A. (1989). Phenotypic plasticity in reproductive traits: evidence from a viviparous snake. *Ecology* 70(6): 1768–1774.

Garner, T. W., Gregory, P. T., McCracken, G. F., Burghardt, G. M., Koop, B. F., McLain, S. E., & Nelson, R. J. (2002). Geographic variation of multiple paternity in the common garter snake (*Thamnophis sirtalis*). *Copeia* 2002: 15–23.

Gillingham, J. C. (1979). Reproductive behavior of the rat snakes of eastern North America, genus *Elaphe*. Copeia 1979: 319–331.

Gloyd, H. K., & Conant, R. (1990). Snakes of the *Agkistrodon* complex: a monographic review. Society for the Study of Amphibians and Reptiles.

Graves, B. M. (1989). Defensive behavior of female prairie rattlesnakes (*Crotalus viridis*) changes after parturition. *Copeia* 1989: 791–794.

Graves, B. M., & Duvall, D. (1993). Reproduction, rookery use, and thermoregulation in free-ranging, pregnant *Crotalus v. viridis*. *Journal of Herpetology* 27(1): 33–41.

Greene, H. W., May, P. G., Hardy Sr., D. L., Sciturro, J. M., & Farrell, T. M. (2002). Parental behavior by vipers. Pp. 179–205. In Biology of the vipers. M. E. Douglas and G. W. Shuett (eds). Eagle Mountain.

Hager, S. B. (2001). The role of nuptial coloration in female *Holbrookia maculata*: evidence for a dual signaling system. *Journal of Herpetology* 35(4): 624–632.

Hall, P. M., & Meier, A. J. (1993). Reproduction and behavior of western mud snakes (*Farancia abacura reinwardtii*) in American alligator nests. *Copeia* 1993: 219–222.

Hoss, S. K., & Clark, R. W. (2014). Mother cottonmouths (*Agkistrodon piscivorus*) alter their antipredator behavior in the presence of neonates. *Ethology* 120(9): 933–941.

Jellen, B. C., & Aldridge, R. D. (2014). It takes two to tango: female movement facilitates male mate location in wild northern watersnakes (*Nerodia sipedon*). *Behaviour* 151(4): 421–434.

Jellen, B. C., Graham, S. P., Aldridge, R. D., & Earley, R. L. (2014). Oestrus in a secretive species: endogenous estradiol varies throughout the shed cycle and influences attractiveness in wild northern watersnakes (*Nerodia sipedon*). *Behaviour* 151(4): 403–419.

King, R. B., Jadin, R. C., Grue, M., & Walley, H. D. (2009). Behavioural correlates with hemipenis morphology in New World natricine snakes. *Biological Journal of the Linnean Society* 98(1): 110–120.

Madsen, T., Shine, R., Loman, J., & Håkansson, T. (1992). Why do female adders copulate so frequently? *Nature* 355: 440–441.

Martin, W. H. (1993). Reproduction of the timber rattlesnake (*Crotalus horridus*) in the Appalachian Mountains. *Journal of Herpetology* 27(2): 133–143.

Martin, W. H. (2002). Life history constraints on the timber rattlesnake (*Crotalus horridus*) at its climatic limits. Pp. 285–306. In Biology of the vipers. M. E. Douglas and G. W. Shuett (eds). Eagle Mountain.

Mason, R. T. (1992). Reptilian pheromones. *Biology of the Reptilia* 18: 114–228.

Means, D. B. (2009). Effects of rattlesnake roundups on the eastern diamondback rattlesnake (*Crotalus adamanteus*). *Herpetological Conservation and Biology* 4(2): 132–141.

Olsson, M., & Madsen, T. (1998). Sexual selection and sperm competition in reptiles. Pp. 503–577. In Sperm competition and sexual selection. T. R. Birkhead and A. P. Møller (eds.). Academic Press.

Plummer, M. V. (1990). Nesting movements, nesting behavior, and nest sites of green snakes (*Opheodrys aestivus*) revealed by radiotelemetry. *Herpetologica* 48(2): 190–195.

Plummer, M. V., & Mills, N. E. (1996). Observations on trailing and mating behaviors in hognose snakes (*Heterodon platirhinos*). *Journal of Herpetology* 30(1): 80–82.

Plummer, M. V., & Snell, H. L. (1988). Nest site selection and water relations of eggs in the snake, *Opheodrys aestivus*. *Copeia* 1988: 58–64.

Rivas, J. A., & Burghardt, G. M. (2005). Snake mating systems, behavior, and evolution: the revisionary implications of recent findings. *Journal of Comparative Psychology* 119(4): 447.

Rossman, D. A. (1996). The garter snakes: evolution and ecology. University of Oklahoma Press.

Saint Girons, H. (1982). Reproductive cycles of male snakes and their relationships with climate and female reproductive cycles. *Herpetologica* 38(1): 5–16.

Schuett, G. W. (1992). Is long-term sperm storage an important component of the reproductive biology of temperate pitvipers. Pp. 169–184. In Biology of the pitvipers. M. E. Douglas and G. W. Shuett (eds). Eagle Mountain.

Schuett, G. W. (1997). Body size and agonistic experience affect dominance and mating success in male copperheads. *Animal Behaviour* 54(1): 213–224.

Schuett, G. W., & Duvall, D. (1996). Head lifting by female copperheads, *Agkistrodon contortrix*, during courtship: potential mate choice. *Animal Behaviour* 51(2): 367–373.

Schuett, G. W., & Grober, M. S. (2000). Post-fight levels of plasma lactate and corticosterone in male copperheads, *Agkistrodon contortrix* (Serpentes, Viperidae): differences between winners and losers. *Physiology and Behavior* 71(3): 335–341.

Schuett, G. W., Harlow, H. J., Rose, J. D., Van Kirk, E. A., & Murdoch, W. J. (1996). Levels of plasma corticosterone and testosterone in male copperheads (*Agkistrodon contortrix*) following staged fights. *Hormones and Behavior* 30(1): 60–68.

Schuett, G. W., Repp, R. A., Amarello, M., & Smith, C. F. (2013). Unlike most vipers, female rattlesnakes (*Crotalus atrox*) continue to hunt and feed throughout pregnancy. *Journal of Zoology* 289(2): 101–110.

Schuett, G. W., Repp, R. A., & Hoss, S. K. (2011). Frequency of reproduction in female western diamond-backed rattlesnakes from the Sonoran Desert of Arizona is variable in individuals: potential role of rainfall and prey densities. *Journal of Zoology* 284(2): 105–113.

Schwartz, J. M., McCracken, G. F., & Burghardt, G. M. (1989). Multiple paternity in wild populations of the garter snake, *Thamnophis sirtalis*. *Behavioral Ecology and Sociobiology* 25(4): 269–273.

Seigel, R. A., & Fitch, H. S. (1984). Ecological patterns of relative clutch mass in snakes. *Oecologia* 61(3): 293–301.

Seigel, R. A., Fitch, H. S., & Ford, N. B. (1986). Variation in relative clutch mass in snakes among and within species. *Herpetologica* 42(2): 179–185.

Seigel, R. A., & Ford, N. B. (1991). Phenotypic plasticity in the reproductive characteristics of an oviparous snake, *Elaphe guttata*: implications for life history studies. *Herpetologica* 47(3): 301–307.

Shaw, C. E. (1948). The male combat "dance" of some Viperid snakes. *Herpetologica* 4(4): 137–145.

Shine, R. (1978). Sexual size dimorphism and male combat in snakes. *Oecologia* 33(3): 269–277.

Shine, R. (2003). Reproductive strategies in snakes. *Proceedings of the Royal Society of London B: Biological Sciences* 270(1519): 995–1004.

Shine, R. (2012). Sex at the snake den: lust, deception and conflict in the mating system of red-sided gartersnakes. *Advances in the Study Behavior* 44: 1–51.

Shine, R. (2014). Evolution of an evolutionary hypothesis: a history of changing ideas about the adaptive significance of viviparity in reptiles. *Journal of Herpetology* 48(2): 147–161.

Shine, R., Olsson, M. M., & Mason, R. T. (2000). Chastity belts in gartersnakes: the functional significance of mating plugs. *Biological Journal of the Linnean Society* 70(3): 377–390.

Taylor, E. N., & DeNardo, D. F. (2005). Reproductive ecology of western diamond-backed rattlesnakes (*Crotalus atrox*) in the Sonoran Desert. *Copeia* 2005: 152–158.

Taylor, E. N., Malawy, M. A., Browning, D. M., Lemar, S. V., & DeNardo, D. F. (2005). Effects of food supplementation on the physiological ecology of female western diamond-backed rattlesnakes (*Crotalus atrox*). *Oecologia* 144(2): 206–213.

Tregenza, T., & Wedell, N. (2002). Polyandrous females avoid costs of inbreeding. *Nature* 415(6867): 71–73.

Uller, T., & Olsson, M. (2008). Multiple paternity in reptiles: patterns and processes. *Molecular Ecology* 17(11): 2566–2580.

Weatherhead, P. J., Barry, F. E., Brown, G. P., & Forbes, M. R. (1995). Sex ratios, mating behavior and sexual size dimorphism of the northern water snake, *Nerodia sipedon*. *Behavioral Ecology and Sociobiology* 36(5): 301–311.

Winne, C. T., Willson, J. D., & Gibbons, J. W. (2006). Income breeding allows an aquatic snake *Seminatrix pygaea* to reproduce normally following prolonged drought-induced aestivation. *Journal of Animal Ecology* 75(6): 1352–1360.

Chapter 6

Anthony, C. D., Venesky, M. D., & Spetz, J. C. (2007). *Eurycea longicauda longicauda* (long-tailed salamander): predation. *Herpetological Review* 38: 175–176.

Barbour, M. A., & Clark, R. W. (2012). Ground squirrel tail-flag displays alter both predatory strike and ambush site selection behaviours of rattlesnakes. *Proceedings of the Royal Society of London B: Biological Sciences* doi:10.1098/rspb.2012.1112.

Birkhead, R. D., Williams, M. I., & S. M. Boback. (2004). The cottonmouth condo: a novel venomous snake transport device. *Herpetological Review* 35: 153–154.

Boback, S. M., Hall, A. E., McCann, K. J., Hayes, A. W., Forrester, J. S., & Zwemer, C. F. (2012). Snake modulates constriction in response to prey's heartbeat. *Biology Letters* 8(3): 473–476.

Boback, S. M., McCann, K. J., Wood, K. A., McNeal, P. M., Blankenship, E. L., & Zwemer, C. F. (2015). Snake constriction rapidly induces circulatory arrest in rats. *Journal of Experimental Biology* 218(14): 2279–2288.

Burghardt, G. M. (1993). The comparative imperative: genetics and ontogeny of chemoreceptive prey responses in natricine snakes. *Brain, Behavior and Evolution* 41(3–5): 138–146.

Chiszar, D., Radcliffe, C. W., Overstreet, R., Poole, T., & Byers, T. (1985). Duration of strike-induced chemosensory searching in cottonmouths (*Agkistrodon piscivorus*) and a test of the hypothesis that striking prey creates a specific search image. *Canadian Journal of Zoology* 63(5): 1057–1061.

Clark, R. W. (2004a). Timber rattlesnakes (*Crotalus horridus*) use chemical cues to select ambush sites. *Journal of Chemical Ecology* 30: 607–617.

Clark, R. W. (2004b). Feeding experience modifies the assessment of ambush sites by the timber rattlesnake, a sit-and-wait predator. *Ethology* 110: 471–483.

Clark, R. W. (2006a). Fixed videography to study predation behavior of an ambush foraging snake, *Crotalus horridus*. *Copeia* 2006: 181–187.

Clark, R. W. (2006b). Post-strike behavior of timber rattlesnakes (*Crotalus horridus*) during natural predation events. *Ethology* 112: 1089–1094.

Clark, R. W. (2016). The hunting and feeding behavior of wild rattlesnakes. Pp. 91–118. In Rattlesnakes of Arizona. G. W. Schuett, M. J. Feldner, C. F. Smith, and R. S. Reiserer (eds.). Eagle Mountain.

Clucas, B., Owings, D. H., & Rowe, M. P. (2008). Donning your enemy's cloak: ground squirrels exploit rattlesnake scent to reduce predation risk. *Proceedings of the Royal Society of London B: Biological Sciences* 275(1636): 847–852.

Crandall, K. A., & Buhay, J. E. (2008). Global diversity of crayfish (Astacidae, Cambaridae, and Parastacidae––Decapoda) in freshwater. *Hydrobiologia* 595(1): 295–301.

Cunningham, D. S., & Burghardt, G. M. (1999). A comparative study of facial grooming after prey ingestion in colubrid snakes. *Ethology* 105(11): 913–936.

Ernst, C. H., & Ernst, E. M. (2003). Snakes of the United States and Canada. Smithsonian Books.

Fleet, R. R., Rudolph, D. C., Camper, J. D., & Niederhofer, J. (2009). Ecological parameters of *Coluber constrictor etheridgei*, with comparisons to other *Coluber constrictor* subspecies. *Southeastern Naturalist* 8(2): 31–40.

Franz, R. (1976). Feeding behavior in the snakes, *Regina alleni* and *Regina rigida*. *Herpetological Review* 7: 82–83.

Gardner, S. A., & Mendelson III, J. R. (2003). Diet of the leaf-nosed snakes, *Phyllorhynchus* (Squamata: Colubridae): squamate-egg specialists. *Southwestern Naturalist* 48(4): 550–556.

Garwood, J. M., & Welsh, H. H. (2005). *Rana cascadae* (Cascades Frog): predation. *Herpetological Review* 36: 165.

Gibbons, J. W., & Dorcas, M. E. (2004). North American watersnakes: a natural history, vol. 8. University of Oklahoma Press.

Glass, J. K. (1972). Feeding behavior of the western shovel-nosed snake, *Chionactis occipitalis klauberi*, with special reference to scorpions. *Southwestern Naturalist* 16(3/4): 445–447.

Gloyd, H. K., & Conant, R. (1990). Snakes of the *Agkistrodon* complex: a monographic review. Society for the Study of Amphibians and Reptiles.

Godley, J. S. (1980). Foraging ecology of the striped swamp snake, *Regina alleni*, in southern Florida. *Ecological Monographs* 50(4): 411–436.

Graham, S. P., Connell, M. R., & Gray, K. M. (2010). *Agkistrodon piscivorus* (cottonmouth): diet. *Herpetological Review* 41: 88–89.

Greene, H. W. (2013). Tracks and shadows: field biology as art. University of California Press.

Greene, H. W., & Rodríguez-Robles, J. A. (2003). Feeding ecology of the California mountain kingsnake, *Lampropeltis zonata* (Colubridae). *Copeia* 2003: 308–314.

Hansknecht, K. A. (2008). Lingual luring by mangrove saltmarsh snakes (*Nerodia clarkii compressicauda*). *Journal of Herpetology* 42(1): 9–15.

Himes, J. G. (2003). Diet composition of *Nerodia sipedon* (Serpentes: Colubridae) and its dietary overlap with, and chemical recognition of *Agkistrodon piscivorus* (Serpentes: Viperidae). *Amphibia-Reptilia* 24(2): 181–188.

Jackson, J. F., & Martin, D. L. (1980). Caudal luring in the dusky pygmy rattlesnake, *Sistrurus miliarius barbouri*. *Copeia* 1980: 926–927.

Kardong, K. V., & Smith, T. L. (2002). Proximate factors involved in rattlesnake predatory behavior: a review. Pp. 253–266. In Biology of the vipers. M. E. Douglas and G. W. Shuett (eds). Eagle Mountain.

Kelehear, C., & S. P. Graham. (2015). *Thamnophis sirtalis* (common gartersnake): diet. *Herpetological Review* 46: 277–278.

Lillywhite, H. B., Sheehy III, C. M., & McCue, M. D. (2002). Scavenging behaviors of cottonmouth snakes at island bird rookeries. *Herpetological Review* 33(4): 259–260.

McCue, M. D., Lillywhite, H. B., & Beaupre, S. J. (2012). Physiological responses to starvation in snakes: low energy specialists. Pp. 103–131. In Comparative physiology of fasting, starvation, and food limitation. M. D. McCue (ed.). Springer.

McFarlane, R. W. (1992). A stillness in the pines: The ecology of the red cockaded woodpecker. Norton.

Messenger, K. R., Shepard, N. A. & Yirka, M. A. (2011). *Plethodon nettingi* (Cheat Mountain salamander): predation. *Herpetological Review* 42: 582–583.

Mizuno, T., & Kojima, Y. (2015). A blindsnake that decapitates its termite prey. *Journal of Zoology* 297(3): 220–224.

Mushinsky, H. R. (1987). Foraging ecology. Pp. 302–334. In Snakes: ecology and evolutionary biology. R. A. Seigel, J. T. Collins, and S. S. Novak (eds.). Macmillan.

Nowak, E. M., Theimer, T. C., & Schuett, G. W. (2008). Functional and numerical responses of predators: where do vipers fit in the traditional paradigms? *Biological Reviews* 83(4): 601–620.

Poran, N. S., Coss, R. G., & Benjamini, E. L. I. (1987). Resistance of California ground squirrels (*Spermophilus beecheyi*) to the venom of the northern Pacific rattlesnake (*Crotalus viridis oreganus*): a study of adaptive variation. *Toxicon* 25(7): 767–777.

Putman, B. J., & Clark, R. W. (2016). Habitat manipulation in hunting rattlesnakes (*Crotalus* sp.). *Southwestern Naturalist* 60: 374–377.

Reichenbach, N. G., & Dalrymple, G. H. (1986). Energy use, life histories, and the evaluation of potential competition in two species of garter snake. *Journal of Herpetology* 20(2): 133–153.

Reinert, H. K., MacGregor, G. A., Esch, M., Bushar, L. M., & Zappalorti, R. T. (2011). Foraging ecology of timber rattlesnakes, *Crotalus horridus*. *Copeia* 2011: 430–442.

Reiserer, R. S. (2002). Stimulus control of caudal luring and other feeding responses: a program for research on visual perception in vipers. Pp. 361–383. In Biology of the vipers. M. E. Douglas and G. W. Shuett (eds). Eagle Mountain.

Reiserer, R. S., & Schuett, G. W. (2008). Aggressive mimicry in neonates of the sidewinder rattlesnake, *Crotalus cerastes* (Serpentes: Viperidae): stimulus control and visual perception of prey luring. *Biological Journal of the Linnean Society* 95(1): 81–91.

Richmond, N. D. (1944). How *Natrix taxispilota* eats the channel catfish. *Copeia* 1944: 254.

Rodríguez-Robles, J. A. (1998). Alternative perspectives on the diet of gopher snakes (*Pituophis catenifer*, Colubridae): literature records versus stomach contents of wild and museum specimens. *Copeia* 1998: 463–466.

Rodríguez-Robles, J. A., Bell, C. J., & Greene, H. W. (1999). Food habits of the glossy snake, *Arizona elegans*, with comparisons to the diet of sympatric long-nosed snakes, *Rhinocheilus lecontei*. *Journal of Herpetology* 33(1): 87–92.

Root, S. T., Carstensen, D., Moore, D., & Ahlers, D. (2014). *Thamnhiphis sirtalis dorsalis* (New Mexico gartersnake): diet. *Herpetological Review* 45: 520.

Rossman, D. A. (1996). The garter snakes: evolution and ecology. University of Oklahoma Press.

Rossman, D. A., & Myer, P. A. (1990). Behavioral and morphological adaptations for snail extraction in the North American brown snakes (genus *Storeria*). *Journal of Herpetology* 24(4): 434–438.

Roth, E. D., May, P. G., & Farrell, T. M. (1999). Pigmy rattlesnakes use frog-derived chemical cues to select foraging sites. *Copeia* 1999: 772–774.

Rundus, A. S., Owings, D. H., Joshi, S. S., Chinn, E., & Giannini, N. (2007). Ground squirrels use an infrared signal to deter rattlesnake predation. *Proceedings of the National Academy of Sciences* 104(36): 14,372–14,376.

Saviola, A. J., Peichoto, M. E., & Mackessy, S. P. (2014). Rear-fanged snake venoms: an untapped source of novel compounds and potential drug leads. *Toxin Reviews* 33(4): 185–201.

Sinclair, A. L., & Lee, J. R. (2014). *Thamnophis sirtalis sirtalis* (eastern gartersnake): diet. *Herpetological Review* 45: 521.

Studenroth, K. R. (1991). *Agkistrodon piscivorus conanti* (Florida cottonmouth): foraging behavior. *Herpetological Review* 22: 60.

Vincent, S. E., Herrel, A., & Irschick, D. J. (2005). Comparisons of aquatic versus terrestrial predatory strikes in the pitviper, *Agkistrodon piscivorus*. *Journal of Experimental Zoology Part A: Comparative Experimental Biology* 303(6): 476–488.

Waters, R. M., & Burghardt, G. M. (2005). The interaction of food motivation and experience in the ontogeny of chemoreception in crayfish snakes. *Animal Behaviour* 69(2): 363–374.

Watkins, J. F., Gehlbach, F. R., & Kroll, J. C. (1969). Attractant-repellent secretions of blind snakes (*Leptotyphlops dulcis*) and their army ant prey (*Neiv amyrmex nigrescens*). *Ecology* 50(6): 1098–1102.

Welsh Jr., H. H., & Lind, A. J. (2000). Evidence of lingual-luring by an aquatic snake. *Journal of Herpetology* 34(1): 67–74.

Williams, B. L., Hanifin, C. T., Brodie Jr., E. D., & Brodie III, E. D. (2012). Predators usurp prey defenses? toxicokinetics of tetrodotoxin in common garter snakes after consumption of rough-skinned newts. *Chemoecology* 22(3): 179–185.

Williams, J., Langshaw, J., & Graham, S. P. (2013). *Ambystoma jeffersonianum* (Jefferson salamander): predation by *Thamnophis sirtalis*. *Herpetological Review* 44: 650.

Williams, M. I., Birkhead, R. D., Moosman, P. R., & Boback, S. (2004). *Agkistrodon piscivorus* (cottonmouth): diet. *Herpetological Review* 35: 271–272.

Chapter 7

Allen, D. C., Morris, D. M., and Vaughn, C. C. (2008). *Nerodia sipedon* (northern watersnake): mortality caused by mussel. *Herpetological Review* 39: 471–472.

Amarello, M. A., & Goode, M. (2004). *Trimorphodon biscutatus* (western lyresnake). *Herpetological Review* 35: 182.

Anderson, J. D. (1956). A blind snake preyed upon by a scorpion. *Herpetologica* 12(4): 327.

Barker, B. S., & Sawyer, Y. E. (2011). *Salvadora hexalepis deserticola* (Big Bend patch-nosed snake): diet and predation. *Herpetological Review* 42: 304.

Beamer, P. D., Mohr, C. O., & Barr, T. R. (1960). Resistance of the opossum to rabies virus. *American Journal of Veterinary Research* 21: 507.

Beane, J. C. (2012). *Heterodon simus* (southern hog-nosed snake): predation. *Herpetological Review* 43: 659–660.

Bennett, A. F. (1994). Exercise performance of reptiles. *Advances in Veterinary Science and Comparative Medicine* 38: 113–113.

Bowman, J., Donovan, D., & Rosatte, R. C. (2006). Numerical response of fishers to synchronous prey dynamics. *Journal of Mammalogy* 87(3): 480–484.

Burger, J., & Zappalorti, R. T. (1986). Nest site selection by pine snakes, *Pituophis melanoleucus*, in the New Jersey pine barrens. *Copeia* 1986: 116–121.

Chabreck, R. H., Holcombe, J. E., Linscombe, R. G., & Kinler, N. E. (1982). Winter foods of river otters from saline and fresh environments in Louisiana. *Proceedings of the Annual Conference of Southeast Associated Fish and Wildlife Agencies* 36: 473–483.

Chapman, J. A., & Feldhamer, G. A. (2003). Wild mammals of North America: biology, management, and economics. Johns Hopkins University Press.

Clark Jr., H. O. (2011). Reptiles and amphibians as loggerhead shrike prey. *Sonoran Herpetologist* 24(3): 20–21.

Conolly, J. W., Beane, J. C., & Edward, C. J. (2015). *Heterodon platirhinos* (eastern hog-nosed snake): predation. *Herpetological Review* 46: 450–451.

Cross, C. L., & Marshall, C. (1998). *Agkistrodon piscivorus piscivorus* (eastern cottonmouth): predation. *Herpetological Review* 29: 43.

Dartez, S. F., Hampton, P. M., & Haertle, N. (2011). *Lampropeltis getula holbrooki* (speckled kingsnake): diet. *Herpetological Review* 42: 292.

Delibes, M., Zapata, S. C., Blázquez, M. C., & Rodríguez-Estrella, R. (1997). Seasonal food habits of bobcats (*Lynx rufus*) in subtropical Baja California Sur, Mexico. *Canadian Journal of Zoology* 75(3): 478–483.

Ellis, D. H., & Brunson, S. (1993). "Tool" use by the red-tailed hawk (*Buteo jamaicensis*). *Journal of Raptor Research* 27(2): 128.

Erickson, D. B. (1978). Robin feeding upon snake. *The Murrelet* 59(1): 26–26.

Ernst, C. H., & Ernst, E. M. (2003). Snakes of the United States and Canada. Smithsonian Books.

Fitch, H. S. (1999). A Kansas snake community: composition and changes over 50 years. Sirsi.

Frye, G. G., & Gerhardt, R. P. (2001). Apparent cooperative hunting in loggerhead shrikes. *Wilson Bulletin* 113(4): 462–464.

Greene, H. W. (2000). Snakes: the evolution of mystery in nature. University of California Press.

Guthrie, J. E. (1932). Snakes versus birds; birds versus snakes. *Wilson Bulletin* 44(2): 88–113.

Hamilton, W. J. (1951). Warm weather foods of the raccoon in New York state. *Journal of Mammalogy* 32: 341–344.

Hartman, C. (1922). A brown rat kills a rattler. *Journal of Mammalogy* 3(2): 116–117.

Holley, J. 2014. Brutal slaying shakes up now-gentle Terlingua. *Houston Chronicle*. April 4.

Hovey, T. E., & Bergen, D. R. (2003). *Rana catesbeiana* (bullfrog): predation. *Herpetological Review* 34: 360–361.

Howell, D. L., & Chapman, B. R. (1998). Prey brought to red-shouldered hawk nests in the Georgia Piedmont. *Journal of Raptor Research* 32: 257–260.

Hummer, J. W., & Tolley, K. (2008). *Thamnophis brachystoma* (short-headed gartersnake): predation. *Herpetological Review* 39: 101–102.

Jones, T. R., & Hegna, R. H. (2005). *Phyllorhynchus decurtatus* (spotted leaf-nosed snake): predator-prey interactions. *Herpetological Review* 36: 70.

Kilmon, J. A. (1976). High tolerance to snake venom by the Virginia opossum, *Didelphis virginiana*. *Toxicon* 14(4): 337–340.

King, K. A., Rorabaugh, J. C., & Humphrey, J. A. (2002). *Rana catebeiana* (bullfrog): diet. *Herpetological Review* 33: 130–131.

Klauber, L. M. (1956). Rattlesnakes, vols. 1–2. University of California Press.

Klenzendorf, S.A., Lee, D. J., & Vaughan, M. R. (2004). *Crotalus horridus* (timber rattlesnake): defense and black bear death. *Herpetological Review* 35: 61–62.

Knight, R. L., & Erickson, A. W. (1976). High incidence of snakes in the diet of nesting red-tailed hawks. *Journal of Raptor Research* 10: 108–111.

Lefranc, N. (1997). Shrikes: a guide to the shrikes of the world. A&C Black.

Livezey, K. B. (2007). Barred owl habitat and prey: a review and synthesis of the literature. *Journal of Raptor Research* 41(3): 177–201.

Lovich, J., Drost, C., Monatesti, A. J., Casper, D., Wood, D. A., & Girard, M. (2010). Reptilian prey of the Sonora mud turtle (*Kinosternon sonoriense*) with comments on saurophagy and ophiophagy in North American turtles. *Southwestern Naturalist* 55(1): 135–138.

Mata-Silva, V., Johnson, J. D., & Barragan, G. (2012). *Sonora semiannulata* (western groundsnake): predation. *Herpetological Review* 43: 661–662.

Mead, J. I., & Van Devender, T. R. (1981). Late Holocene diet of *Bassariscus astutus* in the Grand Canyon, Arizona. *Journal of Mammalogy* 62(2): 439–442.

Meinzer, W., & Sansom, A. (2003). The roadrunner. Texas Tech University Press.

Meyer, K. D., McGehee, S. M., & Collopy, M. W. (2004). Food deliveries at swallow-tailed kite nests in southern Florida. *The Condor* 106(1): 171–176.

Miller, N. 2014. Affidavit highlights evidence in Terlingua murder case. *Odessa American*, February 5.

Moorman, C. E., & Beane, J. C. (2010). *Virginia valeriae valeriae* (eastern smooth earthsnake). *Herpetological Review* 41: 101.

Neal, T., & Steen, D. A. (2015). *Agkistrodon piscivorus* (cottonmouth): predation. *Herpetological Review* 46: 264.

Nelson, S., Kostecke, R. M., & Cimprich, D. A. (2006). *Opheodrys aestivus* (rough green snake): predation. *Herpetological Review* 37: 234.

O'Reilly, R. A. (1949). Shrew preying on ribbon snake. *Journal of Mammalogy* 30: 309.

Parmley, D. (1982). Food items of roadrunners from Palo Pinto County, north central Texas. *Texas Journal of Science* 34: 94–95.

Philpot, V. B., & Smith, R. G. (1950). Neutralization of pit viper venom by king snake serum. *Experimental Biology and Medicine* 74(3): 521–523.

Pough, F. H. (1980). The advantages of ectothermy for tetrapods. *American Naturalist* 115(1): 92–112.

Redmer, M. (1988). Two instances of reptile prey discarded by avian predators. *Bulletin of the Chicago Herpetological Society* 23: 28.

Roble, S. M. (2013). Ophiophagy in red-shouldered hawks (*Buteo lineatus*) with the first record of eastern wormsnakes (*Carphophis amoenus*) as prey. *Banisteria* 41: 80–84.

Rogers, L. L., Mansfield, S. A., Hornby, K., Hornby, S., Debruyn, T. D., Mize, M., & Burghardt, G. M. (2014). Black bear reactions to venomous and non-venomous snakes in eastern North America. *Ethology* 120(7): 641–651.

Rombough, C. J., & Schwab, A. M. (2006). *Rana catesbeiana* (American bullfrog): mortality. *Herpetological Review* 37: 448.

Ross, D. A. (1989). Amphibians and reptiles in the diets of North American raptors. Bureau of Endangered Resources, Wisconsin Department of Natural Resources.

Sherbrooke, W. C., & Westphal, M. F. (2006). Responses of greater roadrunners during attacks on sympatric venomous and nonvenomous snakes. *Southwestern Naturalist* 51(1): 41–47.

Sherrod, S. K. (1978). Diets of North American falconiformes. *Raptor Research* 12(3/4): 49–121.

Smith, J. P. (1997). Nesting season food habits of 4 species of herons and egrets at Lake Okeechobee, Florida. *Colonial Waterbirds* 20(2): 198–220.

Sorrell, G. G. (2004). *Virginia striatula* (rough earth snake): predation. *Herpetological Review* 35: 75–76.

Sperry, C. C. (1933). Autumn food habits of coyotes, a report of progress, 1932. *Journal of Mammalogy* 14(3): 216–220.

Stanback, M. T., & Mercandante, A. N. (2009). Eastern bluebirds provision nestlings with snakes. *North Carolina Academy of Science* 125: 36–37.

Staub, R., Mulks, M., & Merker, G. (2006). *Lampropeltis zonata* (California mountain kingsnake): predation. *Herpetological Review* 37: 231–232.

Steen, D. A., McClure, C. J., Sutton, W. B., Rudolph, D. C., Pierce, J. B., Lee, J. R., &

Guyer, C. (2014). Copperheads are common when kingsnakes are not: relationships between the abundances of a predator and one of their prey. *Herpetologica* 70(1): 69–76.

Stevenson, D. J., Bolt, M. R., Smith, D. J., Enge, K. M., Hyslop, N. L., Norton, T. M., & Dyer, K. J. (2010). Prey records for the eastern indigo snake (*Drymarchon couperi*). *Southeastern Naturalist* 9(1): 1–18.

Strobel, B. N., & Boal, C. W. (2010). Regional variation in diets of breeding red-shouldered hawks. *Wilson Journal of Ornithology* 122(1): 68–74.

Taylor, W. P. (1954). Food habits and notes on life history of the ring-tailed cat in Texas. *Journal of Mammalogy* 35(1): 55–63.

Toweill, D. E., & Teer, J. G. (1977). Food habits of ringtails in the Edwards Plateau region of Texas. *Journal of Mammalogy* 58(4): 660–663.

Weldon, P. J., & McNease, L. (1991). Does the American alligator discriminate between venomous and nonvenomous snake prey? *Herpetologica* 47(4): 403–406.

West, S. K., & Tupacz, E. G. (1985). *Nerodia erythrogaster* (redbelly water snake) and *Buteo lineatus* (red-shouldered hawk). *Catesbeiana* 5: 15.

Winne, C. T., Willson, J. D., Todd, B. D., Andrews, K. M., & Gibbons, J. W. (2007). Enigmatic decline of a protected population of eastern kingsnakes, *Lampropeltis getula*, in South Carolina. *Copeia* 2007: 507–519.

Wood, J. E. (1954). Food habits of furbearers of the upland post oak region in Texas. *Journal of Mammalogy* 35: 406–415.

Wylie, G. D., Casazza, M. L., & Carpenter, M. (2003). Diet of bullfrogs in relation to predation on giant garter snakes at Colusa National Wildlife Refuge. *California Fish and Game* 89: 139–145.

Yacelga, M., & Wiseman, K. D. (2011). *Coluber* (=*Masticophis*) *lateralis euryxanthus* (Alameda Striped Racer) and *Lampropeltis getula californiae* (California kingsnake): predation and maximum prey length ratio. *Herpetological Review* 42: 286–287.

~~~~~~

Chapter 8

Bartram, W. (1958). The travels of William Bartram. University of Georgia Press.

Bond, A. B., & Kamil, A. C. (1998). Apostatic selection by blue jays produces balanced polymorphism in virtual prey. *Nature* 395(6702): 594–596.

Brodie III, E. D., & Janzen, F. J. (1995). Experimental studies of coral snake mimicry: generalized avoidance of ringed snake patterns by free-ranging avian predators. *Functional Ecology* 9(2): 186–190.

Brugger, K. E. (1989). Red-tailed hawk dies with coral snake in talons. *Copeia* 1989: 508–510.

Carpenter, C. C., & Gillingham, J. C. (1975). Postural responses to kingsnakes by crotaline snakes. *Herpetologica* 31(3): 293–302.

Crocol, S. (2002). Live prey in *Buteo* nests. *Kingbird* 52: 137–138.

Do Amaral, J. P. S. (1999). Lip-curling in redbelly snakes (*Storeria occipitomaculata*): functional morphology and ecological significance. *Journal of Zoology* 248(3): 289–293.

Duvall, D., King, M. B., & Gutzwiller, K. J. (1985). Behavioral ecology and ethology of the prairie rattlesnake. *National Geographic Research* 1(1): 80–111.

Ernst, C. H., & Ernst, E. M. (2003). Snakes of the United States and Canada. Smithsonian Books.

Fitch, H. S. (1960). Autecology of the copperhead. University of Kansas Publications of the Museum of Natural History 13: 85–288.

Fitch, H. S. (1965). An ecological study of the garter snake, *Thamnophis sirtalis*. University of Kansas Publications of the Museum of Natural History 18: 495–564.

Fitch, H. S. (1999). A Kansas snake community: composition and changes over 50 years. Kreiger.

Gibbons, J. W., & Dorcas, M. E. (2002). Defensive behavior of cottonmouths (*Agkistrodon piscivorus*) toward humans. *Copeia* 2002: 195–198.

Greene, H. W. (1973). Defensive tail display by snakes and amphisbaenians. *Journal of Herpetology* 7(3): 143–161.

Greene, H. W. (1988). Antipredator mechanisms in reptiles. *Biology of the Reptilia* 16(1): 1–152.

Greene, H. W. (2000). Snakes: the evolution of mystery in nature. University of California Press.

Greene, H. W. (2013). Tracks and shadows: field biology as art. University of California Press.

Greene, H. W., & McDiarmid, R. W. (1981). Coral snake mimicry: does it occur? *Science* 213(4513): 1207–1212.

Heckel, J. O., Sisson, D. C., & Quist, C. F. (1994). Apparent fatal snakebite in three hawks. *Journal of Wildlife Diseases* 30(4): 616–619.

Jackson, D. R., & Franz, R. (1981). Ecology of the eastern coral snake (*Micrurus fulvius*) in northern peninsular Florida. *Herpetologica* 37(4): 213–228.

Jackson, J. F., Ingram III, W., & Campbell, H. W. (1976). The dorsal pigmentation pattern of snakes as an antipredator strategy: a multivariate approach. *American Naturalist* 110(976): 1029–1053.

McCallum, M. L., Trauth, S. E., & Neal, R. G. (2006). Tail-coiling in ringneck snakes: flash display or decoy? *Herpetological Natural History* 10(1): 91–94.

Meshaka, W. E., S. E. Trauth, & C. Files. (1988). *Elaphe obsoleta obsoleta* (black rat snake): antipredator behavior. *Herpetological Review* 19: 84.

Perry, R. W., Brown, R. E., & Rudolph, D. C. (2001). Mutual mortality of great horned owl and southern black racer: a potential risk of raptors preying on snakes. *Wilson Bulletin* 113(3): 345–347.

Pope, C. H. (1937). Snakes alive and how they live. Viking Press.

Pritchett, C. L., & J. M. Alfonzo. (1988). Red-tailed hawk "captured" by striped whipsnake. *Journal of Raptor Research* 22:89.

Rabosky, A. R. D., Cox, C. L., Rabosky, D. L., Title, P. O., Holmes, I. A., Feldman, A., & McGuire, J. A. (2016). Coral snakes predict the evolution of mimicry across New World snakes. *Nature Communications* 7: 11484/doi:10.1038.

Rogers, L. L., Mansfield, S. A., Hornby, K., Hornby, S., Debruyn, T. D., Mize, M., & Burghardt, G. M. (2014). Black bear reactions to venomous and non-venomous snakes in eastern North America. *Ethology* 120(7): 641–651.

Ruane, S. (2015). Using geometric morphometrics for integrative taxonomy: an examination of head shapes of milksnakes (genus *Lampropeltis*). *Zoological Journal of the Linnean Society* 174(2): 394–413.

Ruane, S., Bryson, R. W., Pyron, R. A., & Burbrink, F. T. (2013). Coalescent species delimitation in milksnakes (genus *Lampropeltis*) and impacts on phylogenetic comparative analyses. *Systematic Biology* doi:10.1093/sysbio/syt099.

Shine, R., & Madsen, T. (1994). Sexual dichromatism in snakes of the genus Vipera: a review and a new evolutionary hypothesis. *Journal of Herpetology* 28(1): 114–117.

Sweet, S. S. (1985). Geographic variation, convergent crypsis and mimicry in gopher snakes (*Pituophis melanoleucus*) and western rattlesnakes (*Crotalus viridis*). *Journal of Herpetology* 19(1): 55–67.

Valkonen, J. K., Nokelainen, O., & Mappes, J. (2011). Antipredatory function of head shape for vipers and their mimics. *PloS one* 6(7): e22272.

Vanderpool, R., Malcolm, J., & Hill, M. (2005). *Crotalus atrox* (western diamondback rattlesnake): predation. *Herpetological Review* 36: 191.

Van Heest, R. W., & Hay, J. A. (2000). *Charina bottae* (rubber boa): antipredatory behavior. *Herpetological Review* 31: 177.

Wenner, T. J. (2002). Observation of a gopher snake (*Pituophis catenifer*) constricting a red-tailed hawk (*Buteo jamaicensis*). *Journal of Raptor Research* 46: 323–324.

Williams, B. L., Brodie Jr., E. D., & Brodie III, E. D. (2004). A resistant predator and its toxic prey: persistence of newt toxin leads to poisonous (not venomous) snakes. *Journal of chemical ecology* 30(10): 1901–1919.

Williams, B. L., Hanifin, C. T., Brodie Jr., E. D., & Brodie III, E. D. (2012). Predators usurp prey defenses? toxicokinetics of tetrodotoxin in common garter snakes after consumption of rough-skinned newts. *Chemoecology* 22(3): 179–185.

Yeatman, D. C. (1983). *Virginia v. valeriae* (eastern smooth earth snake): defense. *Herpetological Review* 14: 22.

Young, B. A., Meltzer, K., Marsit, C., & Abishahin, G. (1999). Cloacal popping in snakes. *Journal of Herpetology* 33(4): 557–566.

~~~~~~

Chapter 9

Alirol, E., Sharma, S. K., Bawaskar, H. S., Kuch, U., & Chappuis, F. (2010). Snake bite in South Asia: a review. *PLoS Neglected Tropical Diseases* 4(1): e603.

Blackman, J. R., & Dillon, S. (1992). Venomous snakebite: past, present, and future treatment options. *Journal of the American Board of Family Practice* 5(4): 399–405.

Bush, S. P. (2004). Snakebite suction devices don't remove venom: they just suck. *Annals of Emergency Medicine* 43(2): 187–188.

Ferraro, D. M. (1995). The efficacy of naphthalene and sulfur repellents to cause avoidance behavior in the plains garter snake. Pp. 116–120. In Great Plains wildlife damage control workshop proceedings. R. E. Masters and J. G. Huggins (eds.). Noble Foundation.

Gold, B. S., Barish, R. A., & Dart, R. C. (2004). North American snake envenomation: diagnosis, treatment, and management. *Emergency Medicine Clinics of North America* 22(2): 423–443.

Gutiérrez, J. M., Theakston, R. D. G., & Warrell, D. A. (2006). Confronting the neglected problem of snake bite envenoming: the need for a global partnership. *PLoS Medicine* 3(6): e150.

Gutiérrez, J. M., Williams, D., Fan, H. W., & Warrell, D. A. (2010). Snakebite envenoming from a global perspective: towards an integrated approach. *Toxicon* 56(7): 1223–1235.

Harrison, R. A., Hargreaves, A., Wagstaff, S. C., Faragher, B., & Lalloo, D. G. (2009). Snake envenoming: a disease of poverty. *PLoS Neglected Tropical Diseases* 3(12): e569.

Howarth, D. M., Southee, A. E., & Whyte, I. M. (1993). Lymphatic flow rates and first-aid in simulated peripheral snake or spider envenomation. *Medical Journal of Australia* 161(11–12): 695–700.

Ingraham, C. (2015). The crazy reason it costs $14,000 to treat a snakebite with $14 medicine. *Washington Post.* September 9, https://www.washingtonpost.com/news /wonk/wp/2015/09/09/the-crazy-reason-it-costs–14000–to-treat-a-snakebite- with–14–medicine/?utm_term=.01770f9908c6.

Juckett, G., & Hancox, J. G. (2002). Venomous snakebites in the United States: management review and update. *American Family Physician* 65(7): 1367–1378.

Kasturiratne, A., Wickremasinghe, A. R., de Silva, N., Gunawardena, N. K., Pathmeswaran, A., Premaratna, R., & de Silva, H. J. (2008). The global burden of snakebite: a literature analysis and modelling based on regional estimates of envenoming and deaths. *PLoS Medicine* 5(11): e218.

Langley, R. L. (2005). Animal-related fatalities in the United States—an update. *Wilderness and Environmental Medicine* 16(2): 67–74.

Langley, R. L. (2009). Deaths from reptile bites in the United States, 1979–2004. *Clinical Toxicology* 47(1): 44–47.

Langley, R. L., & Morrow, W. E. (1997). Deaths resulting from animal attacks in the United States. *Wilderness and Environmental Medicine* 8(1): 8–16.

Larréché, S., Mion, G., Mayet, A., Verret, C., Puidupin, M., Benois, A., Petitjeans, F., Libert, N., & Goyffon, M. (2011). Antivenin remains effective against African Viperidae bites despite a delayed treatment. *American Journal of Emergency Medicine* 29(2): 155–161.

Means, B. (2010). Blocked-flight aggressive behavior in snakes. *IRCF Reptiles and Amphibians* 17: 76–78.

Minton, S. A. (1957). Snakebite. *Scientific American* 196: 114–122.

Minton, S. A. (1987). Poisonous snakes and snakebite in the US: a brief review. *Northwest Science* 61: 130–136.

Morgan, B. W., Lee, C., Damiano, L., Whitlow, K., & Geller, R. (2004). Reptile envenomation 20–year mortality as reported by US medical examiners. *Southern Medical Journal* 97(7): 642–645.

National Safety Council. (2015). Injury facts. National Safety Council.

Norris, R. L., Pfalzgraf, R. R., & Laing, G. (2009). Death following coral snake bite in the United States—first documented case (with ELISA confirmation of envenomation) in over 40 years. *Toxicon* 53(6): 693–697.

Parrish, H. M. (1957). Mortality from snakebites, United States, 1950–54. *Public Health Reports* 72(11): 1027.

Parrish, H. M. (1963). Analysis of 460 fatalities from venomous animals in the United States. *American Journal of the Medical Sciences* 245(2): 35–47.

Rakel, R. E. (2000). Saunders manual of medical practice. W. B. Saunders.

Stewart, C. J. (2003). Snake bite in Australia: first aid and envenomation management. *Accident and Emergency Nursing* 11(2): 106–111.

Suchard, J. R., & LoVecchio, F. (1999). Envenomations by rattlesnakes thought to be dead. *New England Journal of Medicine* 340(24): 1930.

Sutherland, S. K., Coulter, A. R., & Harris, R. D. (1979). Rationalisation of first-aid measures for elapid snakebite. *The Lancet* 313(8109): 183–186.

Theakston, R. D. G., & Warrell, D. A. (2000). Crisis in snake antivenom supply for Africa. *The Lancet* 356(9247): 2104.

Waldron, J. L., Bennett, S. H., Welch, S. M., Dorcas, M. E., Lanham, J. D., & Kalinowsky, W. (2006). Habitat specificity and home-range size as attributes of species vulnerability to extinction: a case study using sympatric rattlesnakes. *Animal Conservation* 9(4): 414–420.

Waldron, J. L., Welch, S. M., & Bennett, S. H. (2008). Vegetation structure and the habitat specificity of a declining North American reptile: a remnant of former landscapes. *Biological Conservation* 141(10): 2477–2482.

Williams, D., Gutiérrez, J. M., Harrison, R., Warrell, D. A., White, J., Winkel, K. D., & Gopalakrishnakone, P. (2010). The Global Snake Bite Initiative: an antidote for snake bite. *The Lancet* 375(9708): 89–91.

Wingert, W. A., & Chan, L. (1988). Rattlesnake bites in Southern California and rationale for recommended treatment. *Western Journal of Medicine* 148(1): 37.

WSFA Montgomery. (2013). Rattlesnake bites hunter; hunter uses kit to save his life. WSFA Montgomery. http://www.wsfa.com/story/22083814/rattlesnake-bits -hunter-hunter-uses-kit-to-save-his-life.

~~~~~~

Chapter 10

Crooks, J. A., Soulé, M. E., & Sandlund, O. T. (1999). Lag times in population explosions of invasive species: causes and implications. Pp. 103–125. In Invasive species and biodiversity management. O. T. Sandlund, P. J. Schei, and A. Viken (eds.). Kluwer Academic.

Dorcas, M. E., & Willson, J. D. (2011). Invasive pythons in the United States: ecology of an introduced predator. University of Georgia Press.

Dorcas, M. E., Willson, J. D., Reed, R. N., Snow, R. W., Rochford, M. R., Miller, M. A., & Hart, K. M. (2012). Severe mammal declines coincide with proliferation of invasive Burmese pythons in Everglades National Park. *Proceedings of the National Academy of Sciences* 109(7): 2418–2422.

Fleshler, D. (2015). What a bunch of snakes: the reptile lobby uses the tactics and rhetoric of the gun lobby. *Slate*. December 7, http://www.slate.com/articles/ health_and_science/science/2015/12/the_snake_lobby_defends_dangerous_ invasive_reptile_species.html.

Grimm, F. (2016). Massive fish kill makes Florida water emergency difficult to ignore. *Miami Herald.* March 30, http://www.miamiherald.com/news/local/news-col umns-blogs/fred-grimm/article69081862.html.

Grunwald, M. (2005). Canal may have worsened city's flooding. *Washington Post*. September14,http://www.washingtonpost.com/wp-dyn/content/article/2005/09/13 /AR2005091302196.html.

Grunwald, M. (2006). The swamp: the Everglades, Florida, and the politics of paradise. Simon and Schuster.

Helfman, G., Collette, B. B., Facey, D. E., & Bowen, B. W. (2009). The diversity of fishes: biology, evolution, and ecology. John Wiley & Sons.

Kalthoff, K. (2015). Dallas watches Trinity River levels closely. NBC–5 Dallas. http://www.nbcdfw.com/news/local/Dallas-Watches-Trinity-River-Levels-Closely–303530131.html.

Loftis, R. L. (2010). Trinity River among most polluted waters in Texas. *Dallas Morning News.* January 13, http://www.dallasnews.com/news/texas/2010/01/13/Trinity-River-among-most-polluted-waters–3200.

Mack, R. N., Simberloff, D., Mark Lonsdale, W., Evans, H., Clout, M., & Bazzaz, F. A. (2000). Biotic invasions: causes, epidemiology, global consequences, and control. *Ecological Applications* 10(3): 689–710.

Meshaka, W. E., Butterfield, B. P., & Hauge, J. B. (2004). Exotic amphibians and reptiles of Florida. Krieger.

Meshaka Jr., W. E. (2011). A runaway train in the making: the exotic amphibians, reptiles, turtles, and crocodilians of Florida. Monograph 1. *Herpetological Conservation and Biology* 6: 1–101.

McPhee, J. (1989). The control of nature. Macmillan.

Pimentel, D., Zuniga, R., & Morrison, D. (2005). Update on the environmental and economic costs associated with alien-invasive species in the United States. *Ecological Economics* 52(3): 273–288.

Reed, R. N. (2005). An ecological risk assessment of nonnative boas and pythons as potentially invasive species in the United States. *Risk Analysis* 25(3): 753–766.

Rodda, G. H., Fritts, T. H., McCoid, M. J., & Campbell III, E. W. (1999). An overview of the biology of the brown treesnake (*Boiga irregularis*), a costly introduced pest on Pacific Islands. Pp. 44–80. In Problem snake management. G. H. Rodda (ed.). Cornell University Press.

Romagosa, C. (2009). United States commerce in live vertebrates: patterns and contribution to biological invasions and homogenization (PhD Dissertation, Auburn University).

Romagosa, C. M., Guyer, C., & Wooten, M. C. (2009). Contribution of the live-vertebrate trade toward taxonomic homogenization. *Conservation Biology* 23(4): 1001–1007.

Sweeney, C. (2016). Snake hunt: hunting pythons in Florida with a team of cold-blooded killers. *Popular Science.* http://www.popsci.com/snake-hunt.

Willson, J. D., Dorcas, M. E., & Snow, R. W. (2011). Identifying plausible scenarios for the establishment of invasive Burmese pythons (*Python molurus*) in southern Florida. *Biological Invasions* 13(7): 1493–1504.

Wilonsky, R. (2016). Rawlings hopes $250 million Trinity River park moves from concept to reality by 2021. *Dallas Morning News.* May 20, http://www.dallasnews.com/news/dallas-city-hall/2016/05/20/rawlings-hopes–250–million-trinity-river-park-moves-from-concept-to-reality-by–2021.

~~~~~~

Chapter 11

Abbey, E. (1977). The journey home: some words in defense of the American west. Dutton Books.

Allender, M. C., Baker, S., Wylie, D., Loper, D., Dreslik, M. J., Phillips, C. A., & Driskell, E. A. (2015). Development of snake fungal disease after experimental challenge with ophiodiomyces ophiodiicola in cottonmouths (*Agkistrodon piscivorous*). *PloS One* 10(10): e0140193.

Blehert, D. S., Hicks, A. C., Behr, M., Meteyer, C. U., Berlowski-Zier, B. M., Buckles, E. L., & Okoniewski, J. C. (2009). Bat white-nose syndrome: an emerging fungal pathogen? *Science* 323(5911): 227.

Center for Biological Diversity. (2011). Petition to list the eastern diamondback rattlesnake (*Crotalus adamanteus*) as threatened under the endangered species act. Center for Biological Diversity.

DeLoache, J. S., & LoBue, V. (2009). The narrow fellow in the grass: human infants associate snakes and fear. *Developmental Science* 12(1): 201–207.

De Ruiter, D. J., & Berger, L. R. (2000). Leopards as taphonomic agents in dolomitic caves—implications for bone accumulations in the hominid-bearing deposits of South Africa. *Journal of Archaeological Science* 27(8): 665–684.

Farrer, R. A., Weinert, L. A., Bielby, J., Garner, T. W., Balloux, F., Clare, F., & Anderson, L. (2011). Multiple emergences of genetically diverse amphibian-infecting chytrids include a globalized hypervirulent recombinant lineage. *Proceedings of the National Academy of Sciences* 108(46): 18,732–18,736.

Greene, H. W. (2000). Snakes: the evolution of mystery in nature. University of California Press.

Loureiro, T. G., Anastácio, P. M. S. G., Araujo, P. B., Souty-Grosset, C., & Almerão, M. P. (2015). Red swamp crayfish: biology, ecology and invasion—an overview. *Nauplius* 23(1): 1–19.

Marra, P. P., & Santella, C. (2016). Cat wars: the devastating consequences of a cuddly killer. Princeton University Press.

Peacock, D. (2011). Grizzly years: in search of the American wilderness. Macmillan.

Pimentel, D., Zuniga, R., & Morrison, D. (2005). Update on the environmental and economic costs associated with alien-invasive species in the United States. *Ecological Economics* 52(3): 273–288.

Rosenblum, E. B., James, T. Y., Zamudio, K. R., Poorten, T. J., Ilut, D., Rodriguez, D., & Longcore, J. E. (2013). Complex history of the amphibian-killing chytrid fungus revealed with genome resequencing data. *Proceedings of the National Academy of Sciences* 110(23): 9385–9390.

Schwalbe, C. R., & Rosen, P. C. (1988). Preliminary report on effect of bullfrogs in wetland herpetofaunas in southeastern Arizona. US Forest Service.

Sleeman, J. (2013). Snake fungal disease in the United States. *National Wildlife Health Center Wildlife Health Bulletin* 2: 1–3.

Tschinkel, W. R. (2006). The fire ants. Harvard University Press.

Tuberville, T. D., Bodie, J. R., Jensen, J. B., LaClaire, L. I. N. D. A., & Gibbons, J. W. (2000). Apparent decline of the southern hog-nosed snake, Heterodon simus. *Journal of the Elisha Mitchell Scientific Society* 116(1): 19–40.

Vaillant, J. (2010). The tiger: a true story of vengeance and survival. Vintage.

Wojcik, D. P., Allen, C. R., Brenner, R. J., Forys, E. A., Jouvenaz, D. P., & Lutz, R. S. (2001). Red imported fire ants: impact on biodiversity. *American Entomologist* 47: 16–23.

Index

loser effect, 108
lung, 48
lyresnakes, 43; diet, 139; venom, 127

Madrieta, Albert, 158
male-dominance (mating system), 108, 113
male-male combat, 108–9, 111
malicious killing, 241
Malthus, Thomas, 224
mambas, 20
mammals, as predators, 157–62, 170,
    238–39
mangrove saltmarsh snake, **28**, 138
Marfa, Texas, 151
marine-origins hypothesis, 13–14
Marsh, Othniel C., 14
martens, as predators, 162
Martin, Strother, 67
massasaugas, 31, **87**; basking, 74; as
    endangered species, 252; hibernation,
    95; home range, 92; migration, **87–88**;
    mimicry, 192
maternal care, **121**, 122, **222**
mating: balls, 100–104, 106–9; "dance"
    (*see* male-male combat); and home
    range, 93–94; seasons, 105; systems,
    105–14
May, Peter, **65**, 66
McCarthy, Cormac, 150
Means, Bruce, 210
Meshaka, Walter, 223
metabolism, 50–52
Miami, Florida, 220, 231
Miami Herald, 223
mice, as predators, 158
migration, 87–88
milksnakes: as constrictors, 129; hiber-
    nation, 95; and mimicry, 179, **181**, 182;
    myths, 173
        eastern, 58, 61, 79, 182
        New Mexico, **184**
        red, **181**
mimicry, 22, 179, **180–81**, 192–93
Mississippi River: Control Structure, 220;
    Gulf Outlet Canal, 220
Mitchell, Margaret, 171
Moab, Utah, 2
Mojave Desert, 34–35, 57, 64, 76
Mojave National Preserve, California, 33
moles, as predators, 157
monogamy, 105

Mooney, James, 58
mountain gorillas, 113
mountain lions, 4, 6, 145, 151
mouth gaping, 187, **188–90**; mimicry, 193
movement patterns, 64, 69–71; seasonal,
    88–93
mudsnakes, 19, 43; nests, 114, **115**, 122;
    reproduction, **115**, 119; tail, 49; tail tip,
    186
Muir, John, 233
musk, 180, 183
mussels, 153
myths, 173, 175

Narcisse snake dens, 100–103
National Football League, 237
National Geographic, xi, 224
Natricines: Asian, 19; biology, 17–18;
    reproduction, 117; viviparity, 116
Nerodia, 137
Neslage, Stephen, 228
nests, 114–16
New Mexico Department of Game and
    Fish, 251
nightsnakes, **40**, 151; diet, 139; venom, 127
night vision, **40**, 41
No Country for Old Men (film), 150
nonassociative fears, 242
nondeterministic polynomial time, 231
Nowak, Erika, **244**, 245
Noxen, Pennsylvania, 120

octopus, as predators, 153
Okefenokee Joe, xiii
Okefenokee Swamp, Georgia, 237
olfactory bulb, 37, 41
Olsson, Mats, 100
opossums, as predators, 158
Opp, Alabama, 120
optic tectum, 41
Orianne Society, 248
origin of snakes, 10–15
otters, as predators, 159–60, 162, 170
Outside Magazine, 150
oviparity, 114–16
ovulation, 113–14

Painter, Charlie, 251
palpation, 124
pancreas, 46
Panthera, 246

turtles, as predators, 154
snapping turtles, common, 154
Sonoran mud turtles, 154
Tuskegee National Forest, Alabama, 130, 132

Uktena, 58
University of Georgia, 254
University of Kansas, 174
uric acid, 49
US Fish and Wildlife Service, 251–52
US Special Operations Command, 226

vehicle mortality, 240, **241**
venom, 44; extractors, 216; mild, 43
venomous snakes, **6**; risk, 200
vent, 49
vertebrae, 9, 44
vertebrates, **9**
vestibule, 39
vinesnakes, brown: locomotion, 56; mouth gaping, **190**
Viperidae: biology of, 24–26; distribution of, 6, **8**; fangs, 44; viviparity, **116**
viviparity, 116
vomeronasal organ, 37, 104, 126

Waldron, Jayme, **201**
Walker, Tracy, 254
warning colors, 179, **180**, **181**
Washington, Booker T., 254
watersnakes, 18, 26, 32; basking, reproduction, 117; camouflage, 175; diet, **126**, **137**, 138; head flattening behavior, 192; home range, 89; locomotion, 56; as prey, 153–54; spring emergence, 85
banded, 32; cottonmouth mimicry, 193; diet, **126**, 138; as dietary generalists, **126**; as prey, **159**, **166**
brown, 32; basking, **75**; conservation, 193; cottonmouth mimicry, 193; diet, 137–38; home range, 89; mating, **110**; migration, 88; mortality, 153; mouth gaping, **190**; as prey, 153, 155; reproduction, **110**, 119
common, diet, 132–33
copperbelly, 31, 91; as endangered species, 238, 252
diamondback: diet, 137; musk of, 183
green, 32; diet, **137**
Lake Erie, as endangered species, 252
northern, 31–32; cottonmouth mimicry, 193; home range, 93; mating, 93, **104**, 105–6; reproduction, 117, 119
plain-bellied: diet, 138; mating, **104**
yellowbelly, 32
weasels, 158, 162
Whigham, Georgia rattlesnake roundup, 120, 248, 250
whipsnakes: diet, 139–40; fight with birds of prey, 194; migration, 87; Sonoran, 73; striped, **1**, 5, 95; striped pattern of, 176
whisk ferns, 228
white bursage, 34
white-nose syndrome, 239
white oak runners, 32. *See also* ratsnakes: gray
wilderness, 233–37
Williams, Matt, 130
Williams, Wayne, 172
wolverines, 235
wolves, 145, 151, 251; as predators of snakes, 159
wormsnakes, 43, 64, 201, 216; diet, 133; eastern, reproduction, 119; hibernation, 95; home range, 89–90; as prey, 162

yellow-bellied seasnakes, xii, **23**, 24, 137; as prey, 153; tail, 49; viviparity, 116
yellow jessamine, 99
Yellowstone National Park, 233–36; Yellowstone Falls, Wyoming, 236
YouTube, 157–58

Zappalorti, Bob, 98